Praise for *What We Can Do*

"*What We Can Do* brings cutting-edge sustainability techniques used by chief sustainability officers into the hands and context of everyday individuals. The result? A book that finally empowers readers to maximize their impact across every facet of their lives—from personal day-to-day life choices to professional actions, including investments, policies, and even civic engagement."

—**Ann Tracy**, chief sustainability officer, Colgate-Palmolive

"Remember 'Reduce, Reuse, Recycle'? Drawing on cutting-edge science, *What We Can Do* reimagines classic environmental guidance with a positive and encouraging perspective for the modern era to empower us all to make a difference. A brilliant, needed, and timely book."

—**Michael Martin**, CEO, r.World and former executive director of Concerts for the Environment

"Never has decisive action on climate and environment been more important. This book will help citizens all across America understand that while no one can do everything, everyone can do something, and offers a myriad of ways they can get started too."

—**Dr. Trista Patterson**, Time100 Climate Leader, 2023, and co-founder of Playing for the Planet

"We all hold a beautiful opportunity to activate change and become makers of the future, and *What We Can Do* is a seminal addition to the growing movement that believes in our power to save the world. An era of climate optimism is upon us—are you ready for the journey?"

—**Anne Therese Gennari**, author of *The Climate Optimist Handbook*

"A refreshing reminder of the bipartisan history—and optimistic future—of environmentalism, if we harness our voting power in support of Mother Nature."

—**Craig Shaver III**, former Minnesota state representative (R)

"'Know the impact of thy actions'—this is the mantra of *What We Can Do*, a fast-paced, data-rich, and optimistic examination of the multitude of ways in which we all, individually and collectively, can have positive sustainability impacts."

—**Dr. Tanja Srebotnjak**, executive director of sustainability, Williams College

"*What We Can Do* is a great resource to help readers lead climate positive lives and careers. Importantly, it also provides guidance on one of the most powerful environmental tools we have as citizens: our votes and how to influence the ambitious public policy we need to achieve sustained success."

—**Mark Tercek**, former president and CEO, The Nature Conservancy

"What sets this book apart is Charlie Sellars's refreshing message that every job can be a sustainability job—you don't need to start over or get a new degree. Sellars's practical advice helped me land my dream role; this is the career guide I would recommend to anyone looking to find a career in the climate space."

—**Niraj Poudel**, strategic manager, Watershed

"Climate change impacts us all and the planet we share. Charlie Sellars gives us a refreshing reminder that we are more powerful in making a difference than we think. *What We Can Do* is a must-read for anyone searching for practical guidance on living and working more sustainably."

—**Donna Warton**, corporate vice president, Supply Chain & Sustainability, Microsoft Windows & Devices

WHAT WE CAN DO

A Climate Optimist's Guide to Sustainable Living

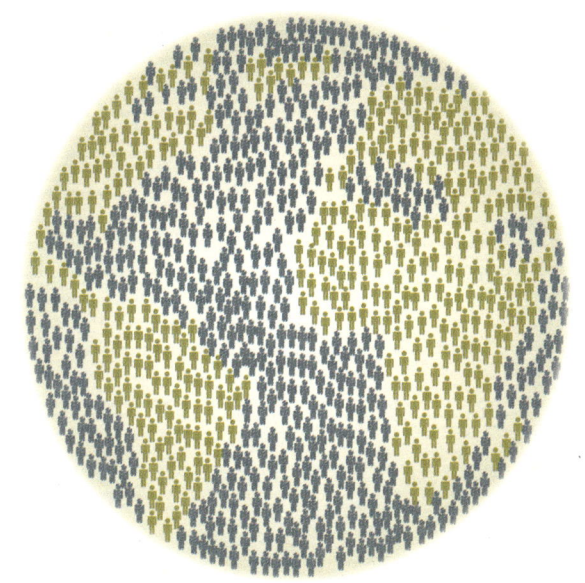

Charlie Sellars

What We Can Do: A Climate Optimist's Guide to Sustainable Living © copyright 2025 by Charlie Sellars. All rights reserved. No part of this book may be reproduced in any form whatsoever, by photography or xerography or by any other means, by broadcast or transmission, by translation into any kind of language, nor by recording electronically or otherwise, without permission in writing from the author, except by a reviewer, who may quote brief passages in critical articles or reviews.

This book represents the personal views and opinions of the author and does not necessarily reflect the positions or opinions of any organization, institution, or individual with which the author is affiliated.

ISBN 13: 978-1-63489-768-6
Library of Congress Catalog Number has been applied for.
Printed in Canada
First Printing: 2025
29 28 27 26 25   5 4 3 2 1

Cover design by Wendy Holdman
Interior design by Ngân Huynh
Author photo by Cadence 3
Edited by Sara Letourneau
Proofread by Caitlin Fultz and Kerry Stapley
Production editing by Lindsay Bohls

Wise Ink
PO Box 580195
Minneapolis, MN 55458-0195
www.WiseInk.com

Wise Ink is a creative publishing agency for game-changers. Wise Ink authors uplift, inspire, and inform, and their titles support building a better and more equitable world. For more information, visit WiseInk.com.

To order, visit www.ItascaBooks.com or www.CharlieSellars.com. Reseller discounts available.

Contact Charlie Sellars at www.CharlieSellars.com for speaking engagements, freelance and consulting projects, and interviews.

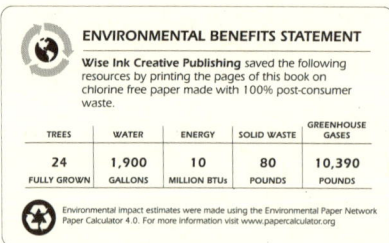 

Based on initial print volume.

Contents

Foreword, by Donna Warton	1
Introduction	5
Part One: Our Personal Lives	23
Chapter 1: Every Decision We Make Has an Impact on Our Planet	25
Chapter 2: Your Family	31
Chapter 3: Your Home (39 Percent of Your "Personal" Emissions)	35
Chapter 4: Your Travel and Transportation (33 Percent of Your "Personal" Emissions)	57
Chapter 5: Your Purchases, Diet, and Stuff (28 Percent of Your "Personal" Emissions)	73
Closing Part One	99
Part Two: Our Professional Lives	103
Chapter 6: Every Job Can Be a Sustainability Job	105
Chapter 7: Where You Work	113
Chapter 8: At Your Work (If You Work a "Normal" Job)	117
Chapter 9: At Your Work (If You Work a "Sustainability" Job)	129
Chapter 10: But What Does Corporate Sustainability Look Like?	135
Chapter 11: Your Investments and Savings	145
Closing Part Two	153
Part Three: Our Political Lives	157
Chapter 12: Our Impact Is Directly Proportional to Our Sphere of Influence	159
Chapter 13: Volunteering and Nonprofits	165
Chapter 14: Sustainability Policies to Support	171
Chapter 15: Sustainability-Adjacent Policies to Support	197
Chapter 16: Engaging Civically and Running for Office	219
Closing Part Three	223
Conclusion	227
Appendix: Select Data and Definitions	235
Glossary	245
About the Author	249

Foreword

It is with great pleasure that I write this foreword for *What We Can Do*, considering Charlie Sellars and I joined forces only four years ago to create a step-function positive change for sustainability within the Windows and Devices business at Microsoft. Always an optimist, Charlie took on the daunting task of connecting product road maps, environmental science, and projects to vision, culture, and ultimately action. Charlie's ability to bring clarity to complex sustainability issues and allow room for others to contribute is a big reason why we have made the progress we have. Even as he, himself, was learning his craft, he created an environment for others to learn as well by bringing experts together to share their insight through the Sustainability Culture of Learning program. His consistent curiosity and optimism in taking on big challenges are a testament to his expertise and a profound insight into his book.

I met Charlie when he was working at Gartner. When a speaker at a Gartner Supply Chain conference he organized talked about the work they were doing on circularity, it left an indelible mark on my thinking. I knew we needed to expand the way we were thinking about sustainability from environmental compliance alone to a holistic approach encompassing our entire operations, from product design to supply chain. My boss at the time saw the potential and supported the effort to create a sustainability program management office. Within the next six months, Microsoft appointed its first chief sustainability officer and made the commitment to be carbon negative, water positive, and zero waste by 2030. It was clear we were onto something transformational. As I looked for talent, I knew I needed someone who could go on the journey with me.

At another Gartner event, in a conversation about the new bold commitments recently announced, Charlie's depth of knowledge on many of the subjects was so impressive that I asked him jokingly, "Why aren't you already working at Microsoft?" He took that as an opening and called me a few weeks later to ask if I was serious about my comment. I shared the thinking behind a new role and the new program management office and that it would be the first of its kind at the company. Charlie's subsequent application stood out, and he has been on the journey ever since. His unique ability to connect with others and make profound ideas accessible defines his approach.

This book is a reflection of Charlie's unparalleled expertise and passion. Each chapter is crafted with meticulous care, offering readers a comprehensive and nuanced understanding of carbon emissions and water conservation both at home and at work. Whether you are a seasoned professional or a curious novice, or just looking for practical advice, this book will be a valuable addition to your library.

What sets this work apart is its practical relevance and actionable insights. Charlie does not merely present theories; he provides a road map for applying these concepts in real-world scenarios. The case studies, anecdotes, and practical examples included in these pages are drawn from his extensive experience and research. They serve as a testament to the practical applicability and transformative potential of the ideas presented.

Moreover, this book is a timely contribution to the climate change discussion. In an era marked by rapid change and unprecedented challenges, Charlie offers a beacon of clarity and direction. His insights are not only relevant but essential for navigating the complexities of our time. This book is more than a collection of ideas; it is a call to action, urging readers to rethink, reimagine, and reinvent their approaches to what they can do to minimize their impact on the planet.

As you embark on this intellectual journey, I encourage you to engage with the material actively. Take notes, reflect on the insights, and consider how you can apply them in your own context. *What We Can Do* is a treasure trove of knowledge, and I am confident that it will leave you enriched and inspired.

In closing, I express my deepest gratitude to Charlie for the unwavering dedication, contributions, and optimism he brings to the climate change issues of today. His work continues to inspire and challenge us, pushing the boundaries of what we know and what we can achieve. It's an honor to introduce this book to you.

—Donna Warton,
corporate vice president,
Supply Chain & Sustainability,
Microsoft Windows & Devices

INTRODUCTION

If you are reading this book, you are (hopefully) already convinced that we, as humans, can become better caretakers of our planet. There are books upon books that talk about the adverse effects of climate change, biodiversity loss, water scarcity, plastic waste, and pollution. This book will not focus any time on proving the need to do better as a human species or fighting "climate denialism." There are other authors far more qualified than me who have already proven the most urgent problem statement of our time.

Instead, this book looks to answer a more personal question:

"What can we do?"

What can we do to fight the climate crisis and keep our planet habitable to us and future generations? What you are about to read will (hopefully) empower you with a more concrete understanding of *what we CAN do* to make a more significant impact across our personal, professional, and political lives. The alternative—being unwilling to change our behavior because we think there is nothing we can individually do—is a form of something I like to call **climate doomism**.

In the eyes of the planet, climate doomism and climate denialism can be equally harmful. To the environment, the only thing that matters is whether the decisions we make as individuals and groups make the planet a better or worse place. And the great news is that all of us—rich or poor, young or old—can do amazing things to make this planet a better place. It is not a matter of doing everything; even I cannot reasonably claim or hope to follow all the recommendations in this book. Any amount of positive impact is good impact, however, and the power of many of us doing a little better far outweighs the power of the small few doing a lot. We cannot let perfect be the enemy of good.

With this initial framing in mind, I hope for this book to inspire **climate optimism** for our potential to make a meaningful difference and to equip us with practical, data-driven methods to enact real change. By the end of this book, I hope to leave you with a framework of thought—backed up by the best data we have available—to help you understand what our biggest impact can be. I hope that this book is both insightful and inspiring rather than obvious and demoralizing. I also hope that you read it not as a mandate for what you MUST do, but rather, as a guide for what you COULD do. That way, we can all make the decisions that fit best into our own lives and realities. None of us can do everything, but all of us can do *something*.

Understanding the Opportunity

I was recently sitting in the audience for a panel on climate change. As I listened to the panelists and the resulting Q&A from the audience that followed, I realized that most people WANT to make a difference, but the average person doesn't really KNOW the most impactful way to make a difference. At best, we have been equipped with broad brushstrokes at the (inter)national policy level (e.g., invest in more renewable energy) and the individual level (e.g., turn off the lights when you leave the room). We are aware of detailed science that shows us how carbon in the atmosphere today compares to millions of years ago and how biodiversity is being adversely impacted by deforestation and plastic waste. We also know the big-ticket items that governments must pursue and the deadline for when to curtail some of the worst impacts of climate change.

But curiously enough, there are almost no frameworks that help us as *individuals* to understand what we can do to maximize our impact across everything we do. We can certainly intuit some things to be more sustainable than others, but how can we actually know? How can we be sure we're doing the best things for the planet—and not accidentally spending our time and effort on things that, at best, don't help that much and, at worst, may be harmful?

Per a 2023 study of over one thousand Americans conducted by Yale titled "Global Warming's Six Americas," while the majority of Americans agree that climate change is driven by human activity and are concerned about its effects, most of that majority—ostensibly, the ones most educated and interested in the topic—aren't even sure what they can or should do about it.[1] (Thankfully, reading this book should help demystify this question!)

---

[1] "Global Warming's Six Americas," Yale Program on Climate Change Communication, Yale University, last modified June 11, 2024, https://climatecommunication.yale.edu/about/projects/global-warmings-six-americas/.

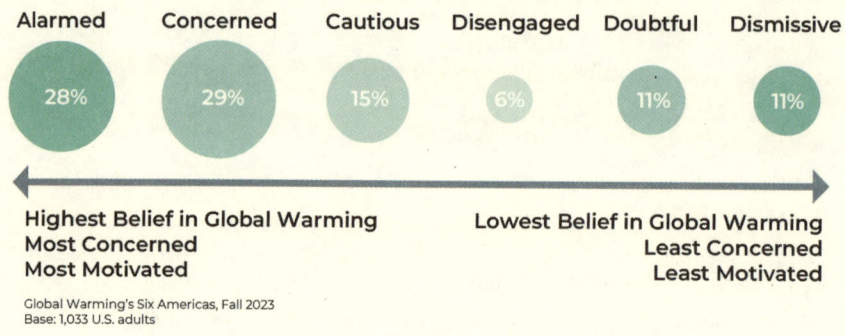

Source: "Global Warming's Six Americas," Yale Program on Climate Change Communication, https://climatecommunication.yale.edu/about/projects/global-warmings-six-americas/.

Here's one critical reason why it's so hard to know what we can do to make the world a better place: **We are at day one—maybe even day zero—of learning how to measure the impact of our individual decisions**. It turns out that figuring out how to calculate the impact of things is *really* hard. Every single decision that we make has a sustainability implication, and each of those decisions is intertwined with countless other impact-laden decisions that are nearly impossible to track. The field of study to track this—**carbon accounting**, writ large—is still nascent.

Let me give an example. As the conscious consumer, let's say you're in the grocery store, trying to figure out which of two deodorant brands is more sustainable. Here's a list of just some of the things that are critical to answering that question that you (and perhaps even the manufacturer) probably don't have access to:

- How sustainable is each material that went into the product and its packaging?
- Where did each brand source its raw materials from? And how far away is each source? Also, how did those raw materials get to the manufacturer? By airplane? Boat? Train?
- What kind of energy powered the extraction and processing of those raw materials? Is the manufacturing facility powered by dirty energy? Or is the building energy-efficient?
- How far did the brand's employees have to commute to the manufacturing plant?
- Were fair labor standards applied across every layer of that manufacturing chain?
- How far did the product travel to get to the retailer? Were the truck drivers driving fuel-efficient vehicles? Were they driving efficiently?
- How energy-efficient is the store you're buying the deodorant from? Is it more sustainable to order online or to drive to the store?
- What kind of "emissions overhead" is needed to run the store and the brand (e.g., employees' computers, use of cloud storage, energy to power the headquarters)?

The list, unfortunately, goes on and on. But it's not just a challenge for the customer. The companies making the products are also learning how to answer all these questions.

I'll give you an example of how tricky it can be to calculate the impact of a product based on my experience at Microsoft as a director of sustainability. For our consumer hardware, we measure its impact and footprint using something called a **lifecycle assessment**, which attempts to take a bottom-up, holistic view into calculating the impact of products. This emerging technique looks at the full impact of *every single part that goes into a product*, all the way to the raw materials that formulated them and all the processes to smelt and refine them by hundreds of different suppliers. It also incorporates the real-time transportation of the product to the store and the usage of the product by our customers over its lifetime.

When calculating this for a single product like a PC, a huge number of data fields need to be tracked across the entire supply chain and captured against hundreds of different parts. In many cases, we have to make assumptions and estimations where primary data doesn't exist. At an average company, conducting these lifecycle assessments usually requires a number of highly technical experts (who likely aren't on staff) spending dozens of hours (or more) assembling a completed view for just a single product. Scale this work across a full portfolio of products that may each need their own analysis, and you can understand how this can get really complex really quickly—and why many companies haven't yet figured out how to do it.

Historically, we've been better at understanding things from a top-down or ex post facto point of view, such as by tracking the amount of garbage generated annually and collected by municipalities, measuring the carbon dioxide growth in our atmosphere, or measuring the size and growth of the Great Pacific Garbage Patch. Going a step further, the United States Environmental Protection Agency (EPA) published a 919-page *Inventory of U.S. Greenhouse Gas Emissions and Sinks* in 2024. It revealed that out of America's roughly 6.3 billion tons of gross greenhouse gas emissions in 2022, 30 percent came from industry, 29 percent from transportation, 16 percent from commercial buildings, 15 percent from residential, and 10 percent from agriculture.[2] (Note that emissions from electricity generation are allocated to the above sectors based on their energy usage.) At a high level, we've even started figuring out how to allocate out "who did what" to contribute under these country-level figures, hence why you may have heard the oft-quoted (and misused) statistic that "71 percent of global industrial emissions come from just 100 companies."[3]

However, that view—and the philosophical weight behind it—is limited, especially when it comes to helping us everyday people understand what we can do to make more of a difference. This is where the growing understanding of bottom-up impact, like the life cycle assessment method mentioned earlier, is becoming a powerful tool to translate the knowledge we do have into everyday decisions, even if the data is imperfect or not fully available. Already, the insights we've received from this practice have given us the right directional framework on how to understand and mitigate the impact, even if the numbers continue to evolve.

While we might have previously felt powerless when we heard (misleadingly) that one hundred companies contribute 71 percent of all global industrial emissions, the bottom-up approach helps highlight that **we are actually a part of the solution**. An enormous part of ExxonMobil's emissions, for example, includes counting the fuel that we use to drive our cars, the plastic

---

2 United States Environmental Protection Agency, *Inventory of U.S. Greenhouse Gas Emissions and Sinks*, 2024, https://www.epa.gov/system/files/documents/2024-04/us-ghg-inventory-2024-main-text_04-18-2024.pdf.

3 Carbon Disclosure Project, "New Report Shows Just 100 Companies Are Source of Over 70% of Emissions," July 10, 2017, https://www.cdp.net/en/articles/media/new-report-shows-just-100-companies-are-source-of-over-70-of-emissions.

WHAT WE CAN DO

packaging that we use and discard, and the polyester in the clothes that we wear.[4] (Yeah, oil is kind of in everything that we use today.)

These are all decisions we make every day that are well within our control to make better. We are far from powerless. Per ExxonMobil's 2023 greenhouse gas emissions report, the company estimated that it emitted between 97 and 110 million tons of carbon dioxide emissions from its operations to acquire and refine fuel—and that customers burning its fuel produced between 540 to 720 million metric tons of emissions.[5] This means that all of us combined potentially contribute upward of 80 percent of ExxonMobil's emissions. And while that doesn't mean that a company like ExxonMobil gets to wash its hands of its emissions (it absolutely has the lion's share of work to do), it does start to turn the page on us feeling helpless by these doomsday statistics built on top-down analysis, and gives us a bit more hope on the power we, as individuals, can have in bottom-up impact.*

*It may be worth noting that the suggestions in this book will only touch on things that are concordant with modern society and will avoid discussion of grossly unethical or "extrajudicial" considerations. For example, if you are looking for advice on how to sabotage an oil rig as a way to have a bottom-up impact, I would recommend seeking out other literature.

---

[4] ExxonMobil, *Advancing Climate Solutions: Progress Report, GHG Data Supplement*, April 2023, https://corporate.exxonmobil.com/-/media/global/files/advancing-climate-solutions-progress-report/2023/2023-acs-ghg-data-supplement.pdf.

[5] ExxonMobil, *Advancing Climate Solutions: Progress Report*.

The Framework and Measurement

So given all of this context, what does that mean for us? How can we take advantage of this burgeoning—and imperfect—field of bottom-up impact to meaningfully guide our decisions?

Before getting into the framework for harnessing the power of this analysis, I want to quickly stress that **strict numeric accuracy is not the point of this framework**. Because these methods for capturing impact are so complex and new, we are constantly evolving our understanding of how to quantify impact. How we measure things can (and will) change over time, as will the complex systems that support and produce the goods and infrastructure in our lives. It could even evolve so much that it changes some of the guidance in this book!

To combat this and future-proof the book to the best of my capability, most of the data sources in this book will be pulled directly from as many reputable and "official" sources as possible, the most frequent ones being the Energy Information Administration (EIA), the EPA, the Congressional Budget Office, and the United States Geological Survey (USGS). But even then, sometimes this data is years old—if it exists at all. Being educated in sustainability requires understanding that our knowledge can continually evolve and improve, like with any other science.

Instead, **consider this framework a guidepost to understand the relative impact of things as we understand them today**, such as orders of magnitude and categories of impact. It is the best we have today, and it's still immensely powerful to make the "right" impact decision more often than not. We know enough now to make most of our impact decisions with a pretty high degree of confidence.

With all that being said, here are several themes and frameworks that we will explore over the course of this book:

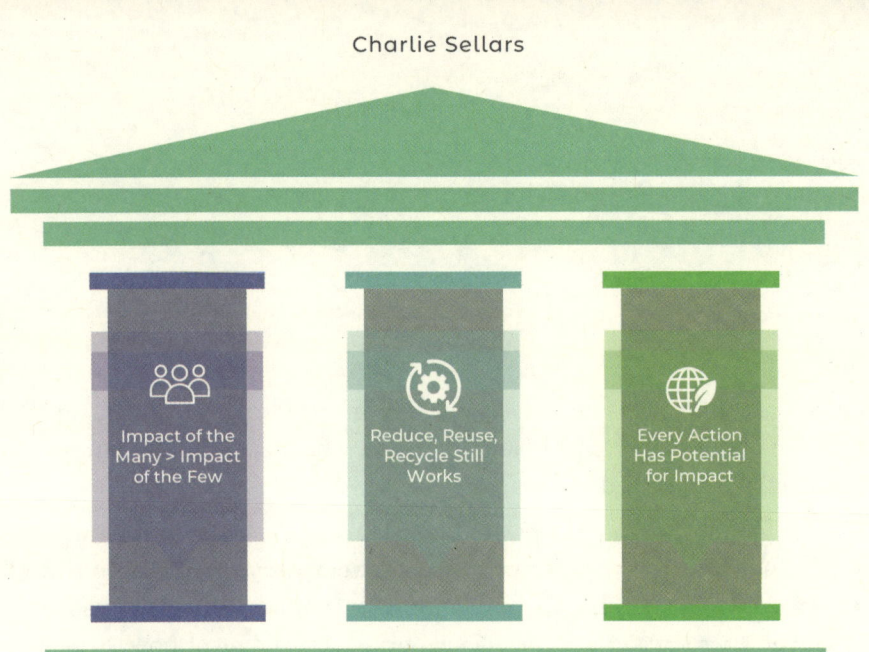

- **The impact of the many exceeds the impact of the few.** While this impact is hopefully intuitive, I am constantly surprised at how often people's conception of their ability to help the planet ends with themselves and their individual actions. Luckily, as individuals, we are also contributors to much larger groups of individuals, such as the nonprofits we volunteer with, the worship or study groups we belong to, the jobs we work at (and their customers), and the governments and communities that we reside within. Depending on which groups we are actively part of, we can go from an "impact of one" to an impact of hundreds, thousands, or even millions by leveraging the social fabric we're a part of. The rest of this book will be organized into parts that follow this structure of exponentially growing impact potential, from the things you can do for yourself and your family (the "Personal"), to ways to drive sustainability in your career and work (the "Professional"), to how best to engage your community and government for collective impact (the "Political").

- **"Reduce, Reuse, Recycle" is still the most effective framework to minimize our negative impact on the planet.** This is due, in part, to a prevailing truth that has emerged as our sustainability science and our understanding of lifecycle assessments have evolved. First, the emissions, water, and waste to MAKE the things you own (also known as **embodied carbon** for emissions specifically) are major contributors to the overall emissions, water usage, and waste of a product—and this impact often exceeds using or consuming those things. Nearly everything we buy will follow this pattern. If you're reading this book on a tablet or e-reader, for example, the emissions generated to make that machine can be almost ten times as much as the emissions generated to use it over its life. The more we can reduce our consumption as a first step, the more we can avoid the waste to make and use. Plus, the more we can reuse (and repair) our things as a second step, the more we can reduce our waste and throw away less stuff. And though the efficacy of recycling has been called into question of late, the ability to recycle *some* of our things and give them new life still outweighs throwing them away outright.

- **You have the power to make a sustainable impact in everything you do.** And I do mean *everything*. Every decision we make—in our daily lives, at our jobs, at the ballot box—has an impact tied explicitly or implicitly to it. Your job has the power to be a sustainability job. Your community has the power to become a climate-friendly community. Your vote has the power to be a sustainability vote. Rather than being debilitating, this should be exhilarating. Once you can see the impact of your decisions (hopefully by using the learnings in this book!), you'll see how this truth can be an empowering one—and a counter to climate doomism and the sense that there is "nothing we can do."

Depending on the topic, I'll also introduce additional concepts such as "per capita" thinking (i.e., dividing the emissions of an activity by the number of people serviced by that activity). For example, your car will produce the same amount of absolute emissions driving to work whether you drive yourself or carpool. But ideally, carpooling can help your coworker avoid having to drive themselves to work and generate more emissions from their own car. To get the truest sense of the emissions that each person in that car is accountable for, taking a per-capita approach (i.e., dividing the total emissions from gas used by the number of people in the car) is best. This will become especially important when thinking of household emissions.

Before diving into these three categories, another thread that I hope to pull through this book is to make our impact more "real" as I walk through the various things we can do to make the world a better place. As of 2024, there is an excellent—if high-level and directional—Greenhouse Gas Equivalencies calculator[6] managed by the EPA that will be used to help convert more esoteric

---

[6] "Greenhouse Gas Equivalencies Calculator," United States Environmental Protection Agency, last modified January 2024, https://www.epa.gov/energy/greenhouse-gas-equivalencies-calculator. Following equivalencies rely on this tool if no additional footnote is listed.

terms in sustainability (e.g., metric tons of carbon dioxide equivalent, thousands of cubic feet of natural gas) into more tangible, relatable comparisons such as "number of trees" or "miles driven." Even though the specific numbers and comparisons are likely to change and evolve as our knowledge and measurement of sustainability evolves, understanding the relative order of "impact magnitude" can help give us context. (I have also provided a glossary at the end of this book to define some of the terms you'll come across here.)

Not being able to see and understand the direct impact of our actions is, sadly, a consequence of our modern society. **A lot of the real "impact" of our activity is made invisible to us.** We are (normally) not there when our meat is killed, our jewelry is mined, our forests are cut down to build our furniture, or an oil well is drilled to produce the plastic in our potato chip bags. Further, our society has intentionally built structures to make our impact invisible. For instance, municipal waste services—a key part of maintaining a livable society—also obscure the true amount of waste we produce. In 2018, the EPA estimated that the average American was generating around 5 pounds of waste every single day, of which only around 30 percent is recycled or composted.[7] If you are an average American, each year you generate over 1,200 pounds of *pure trash*—much of which does not biodegrade and will outlast even your grandchildren.

And this is just the waste of the end products that you use and consume. There's often even *more* waste generated upstream during the construction, creation, and distribution of these goods to get to your door. As an example, from a greenhouse gas emissions perspective, the following activities are roughly the equivalent of burning down a ten-year-old medium-growth tree:

- Eating twenty burgers[8]

---

7 "National Overview: Facts and Figures on Materials, Wastes and Recycling," United States Environmental Protection Agency, last modified on November 22, 2023, https://www.epa.gov/facts-and-figures-about-materials-waste-and-recycling/national-overview-facts-and-figures-materials.

8 Stephen Ferguson, "Engineering the Low-Carbon, Cruelty-Free, Lab-Grown Hamburger of the Future," Siemens, February 4, 2022, https://blogs.sw.siemens.com/simcenter/engineering-the-low-carbon-lab-grown-hamburger-of-the-future/.

- Buying and wearing two pairs of jeans[9]
- Using a week's worth of electricity in the average American home[10]
- Heating a Midwestern home for three days in the winter[11]
- Driving 180 miles solo in an average American vehicle[12]
- Your portion of one flight from San Francisco to Los Angeles[13]

Some things in our lives create even more impact than that, like the emissions from manufacturing your computer[14] (roughly three trees) or making a single Bitcoin transaction[15] (over six trees). And once you start layering in the amount of water it requires to do these things, the impact becomes even more stark. While two pairs of jeans are the emissions equivalent of burning down a tree, it is the water equivalent of how much you would drink over *seven years*. And eating those twenty burgers is the equivalent of drinking over fifty years' worth of water![16]

How on earth could that be possible, you ask? It underscores the second principle of this book: **Making things takes an unbelievable number of resources**. Cows (and their food) require a LOT of water to grow well before reaching your plate. Your pair of jeans requires a LOT of water-thirsty cotton to make—as well as even more water, electricity, and heat to wash and dry them.

At first, some of this data may be jarring. Some of it may even be shocking or confusing. But by the end of this book, it should be empowering.

---

9 Levi Strauss & Co., *The Life Cycle of a Jean: Understanding the Environmental Impact of a Pair of Levi's ® 501® Jeans*, 2015, https://levistrauss.com/wp-content/uploads/2015/03/Full-LCA-Results-Deck-FINAL.pdf.

10 "Use of Energy Explained," US Energy Information Administration, last modified December 18, 2023, https://www.eia.gov/energyexplained/use-of-energy/homes.php.

11 US Energy Information Administration, "Table WF01. Average Consumer Prices and Expenditures for Heating Fuels during the Winter," October 2022, https://www.eia.gov/outlooks/steo/tables/pdf/wf-table.pdf.

12 United States Environmental Protection Agency, "Greenhouse Gas Equivalencies Calculator."

13 "Google Flights," Google, accessed July 2024, https://www.google.com/travel/flights/.

14 "Eco Profiles," Microsoft, accessed July 2024, https://www.microsoft.com/en-us/download/details.aspx?id=55974.

15 "Bitcoin Energy Consumption Index," Digiconomist, accessed July 2024, https://digiconomist.net/bitcoin-energy-consumption.

16 "The Water Content of Things: How Much Water Does It Take to Grow a Hamburger?," United States Geological Survey, accessed July 2024, https://water.usgs.gov/edu/activity-watercontent.php. Combined with adult water consumption data, "Fast Facts: Data on Water Consumption", US Center for Disease Control, last modified January 19, 2024, https://www.cdc.gov/nutrition/php/data-research/fast-facts-water-consumption.html.

Our First Meta-Analysis: What About the Way You're Reading This Book?

To give a flavor of how we'll explore the above themes in practice, let's take a moment to get meta: Realizing it may be too late if you're already reading this, what is the most sustainable way in which you can enjoy this book and minimize the waste, water, and carbon footprint of reading it?

- **The impact of the many exceeds the impact of the few:** Assuming that this book can help individuals become more sustainable, the best thing you can do is have your friends, families, and coworkers read your copy (or get their own). The positive impact that can ideally be generated from this text hopefully outweighs the emissions of writing and distributing the book itself.

- **"Reduce, Reuse, Recycle" is our most effective framework:** If possible, you shouldn't buy a physical copy of this book. Instead, if you can get it used (such as from the library, a used bookstore, or a friend), that will help distribute the environmental cost of making the book over more people. If you already own a digital e-reader, buying this book digitally is a great choice. And if you simply *must* own a new copy, we have done what we can to minimize the impact of making this book: We are *reducing* by eliminating plastic wrap and using inks that are partially vegetable-based, and *reusing* by printing on 100% post-consumer waste (PCW) recycled paper. We're also improving the book's own ability to be reused and shared over the years by using more durable thread-sewn binding. To top it off, we have committed to donate a portion of the book's proceeds to environmental causes through its 1% for the Planet certification.

- **You have the power to make a sustainable impact in *everything* you do:** Ultimately, the best way to ensure that this book is worth the paper and ink it took to make it is to do something with it. While it's impossible to try everything listed in this book, there's so much that we can do if we only know to think and look for it. I'll consider this book an environmental success if each reader finds themselves taking on at least one behavioral change.

And with that, let's jump into the book!

> **Discussion Points**
>
> 1) Do you personally feel aspects of climate doomism? What do you think are the main drivers of this doomism in yourself or others? Does this inhibit or enhance your desire to live a more sustainable life?
>
> 2) Are there any parts of this book's framing that surprised you? Why?
>
> 3) What do you hope to get out of this book?

PART ONE: OUR PERSONAL LIVES

Chapter 1: Every Decision We Make Has an Impact on Our Planet

While some decisions are big and others are small, their impact can add up in a material way over an average person's lifetime. While that might seem overwhelming, it is also empowering in that we can do better for the planet every single day.

These decisions we make in our personal lives are numerous and varied, from "What brand of jeans should I buy?" to "What should I set the heat to?" to "Should I move to the suburbs?" Unfortunately, the vast majority (if not all) of these decisions exist in a climate impact vacuum. There is simply not enough maturity in, or availability of, impact assessments today to accompany the decisions we make with quantitative impact. Certain things might seem obvious (e.g., turning off the lights when you leave a room), but how much do the decisions we make every day matter? In the absence of data, how can we build a mental model for how to gravitate toward the more sustainable choice?

If you look at an "average" American—with an average house, living in a household of 2.5 people, one car, a dog, and regular diet and shopping behaviors—on an annualized basis, this is how our "personal life" emissions break down, excluding the impact of things we do at our jobs or in our communities, which I will layer in later in the book. (Supporting calculations and sources can be found in the appendix.)*

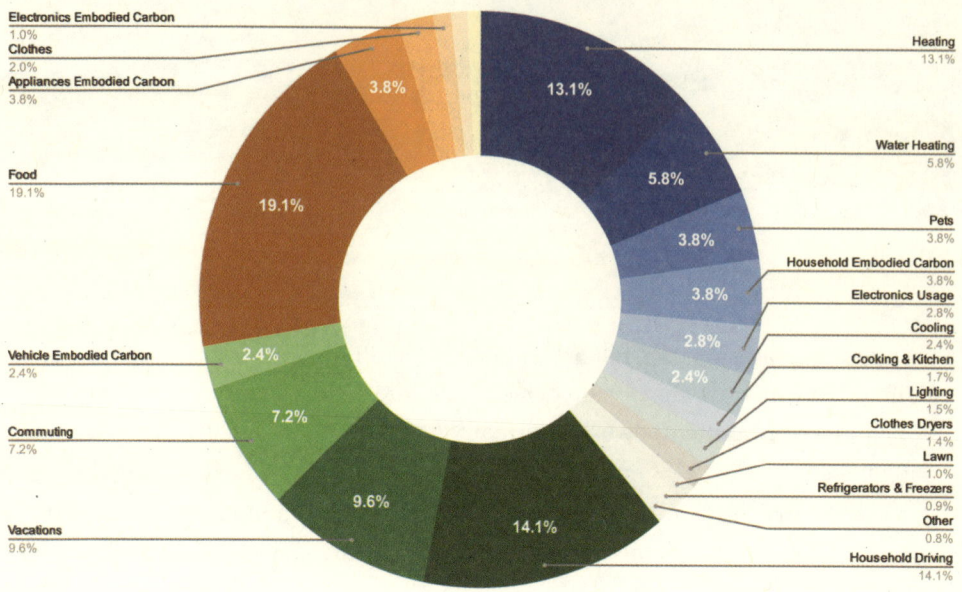

*This view can change considerably depending on various factors, including the number of cars owned, the type of diet, the number of people living in your household, the age of inhabitants, and the types of vacations (if any) taken. It apportions out per-capita emissions between theoretical household members for certain items that are assumed to be shared. It also omits certain types of emissions like services rendered, health care, and common infrastructure such as roads and buildings. Consider this view directional and not absolute.

As I pulled this view together for the book, compiling data across many different sources, it amazed me to learn of the relative impact of certain things versus others. Personally, I had completely underestimated the emissions impact of using hot water and overestimated the emissions impact of air conditioning (AC). For an average American, around 39 percent of our annual personal emissions come from our homes (e.g., heating, cooling, electricity, cooking), 33 percent from our transportation (e.g., commuting, driving, vacationing), and the remaining 28 percent from our purchases (e.g., food, clothes, electronics, furniture). Across all categories, heating our homes and water, driving, and our diets are, by and large, the biggest sources of emissions each year. For some of us with unaverage lifestyles, things like "traveling for business" (which is not listed on the graph) could end up contributing a significant chunk of emissions. I encourage you, the reader, to reflect on your life and think about where your lifestyle choices may affect your emissions pie chart.

WHAT WE CAN DO

"Make It, Move It, Use It, Lose It"

So what can we do about all this? Thankfully, many of us have been equipped with a simple but incredibly insightful mnemonic and hierarchy of decision-making for most of our lives: "Reduce, Reuse, Recycle." But from a data-driven perspective, this is a profound archetype for how we can make more sustainable decisions.

To elaborate, I'd like to give a brief overview as to where emissions and waste *actually* come from. Spoiler alert: **We need to always ask ourselves, "What was the impact of MAKING something in addition to USING it?"**

Take a computer, for example. It is perhaps intuitive to think that the majority of your impact can come from switching on a more energy-friendly default setting or turning it off when you're not using it. While these things are certainly helpful, that computer—and all of the raw materials used to put it together—have taken a circuitous, often impact-laden pathway to get into your home well before you even turn it on for the first time.

In reality, a huge chunk of the emissions associated with most products tend to come from making the product in the first place and getting it to your doorstep. Simplified, the impact of your average product is broken into four categories:

1) **"Make It"**: This refers to the assembly of the product from all of its components.

2) **"Move It"**: This refers to how the product gets to you as the customer.

3) **"Use It"**: This refers to the energy you as the customer burn to use the product.

4) **"Lose It"**: This refers to the energy required to recycle, incinerate, or throw away the product at the end of its life.

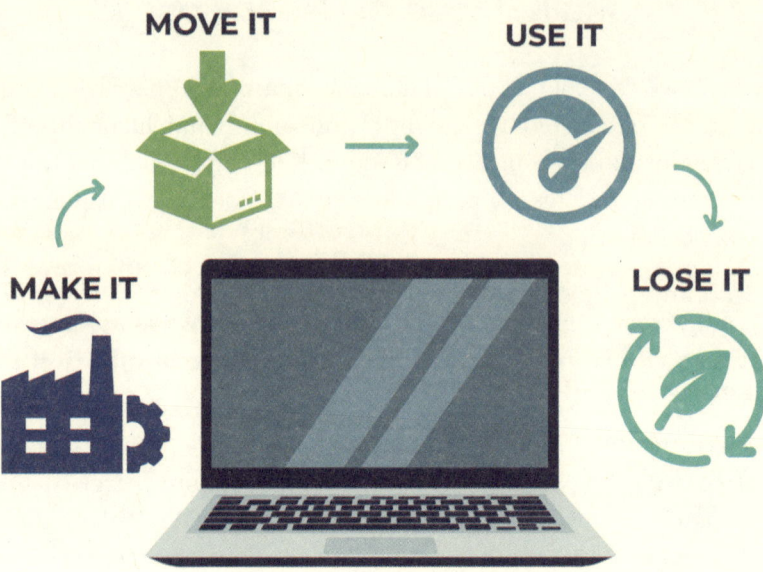

The Framework in Action

For the average laptop, the majority of the lifetime emissions of the product come from making it in the first place.[17] Think about the energy involved in making every single thing in your computer. The aluminum in the enclosure, the cobalt in the battery, and the gold in the circuit boards all likely came from mines, which then sent the raw materials to smelters, which then sent them to component builders, which then sent them to energy-intensive PC assemblers, which then sent them to us.

When tracing the full supply chain of the hundreds of individual components that go into a computer, it is then perhaps understandable that this impact would exceed that of using the device. Even for an Apple MacBook Pro, which uses low-carbon recycled materials and renewable energy in portions of its supply chain, the manufacture of the device is 80 percent of its lifetime emissions.[18] You would need to use that MacBook Pro for over a decade before the emissions from charging your laptop would exceed the emissions needed to make the device in the first place!

This is true for water as well. According to the United Nations[19] and the

---

17 "Eco Profiles," Microsoft. I used this as an example of average laptop emissions breakdowns.

18 Apple, *Product Environmental Report: 14-Inch MacBook Pro*, 2023, https://www.apple.com/environment/pdf/products/notebooks/14-inch_MacBook_Pro_PER_Oct2023.pdf.

19 Johnny Wood, "Your Morning Cup of Coffee Contains 140 Litres of Water," World Economic Forum, March 22, 2019, https://www.weforum.org/agenda/2019/03/hidden-water-in-your-cup-of-coffee/.

Water Footprint Network,[20] it can take thirty-five *gallons* of water to produce a single cup of coffee. You could drink over one thousand(!) cups of water before using the same amount of water as your morning brew. How is this possible? It turns out that it takes a lot of water to grow coffee beans and create the raw materials needed to process, package, and ship the beans to your local store.

Generally, except for highly energy-intensive or very long-lasting products, this general rule holds true: **Almost everything you engage with in your daily life takes more energy, water, and waste to make than it will for you to use it**. One day this could change—once our economy can convert its supply chains and materials to have net-zero impact, that is—but that day is likely decades away.

Depending on the item, moving it to get to you can have a significant impact as well. Especially for items that are heavy or are relatively simple to make, getting the product to your door can be an important factor. For example, per IKEA's 2023 sustainability report, moving its furniture products (both to the store and to our homes) represents over 10 percent of IKEA's emissions.[21] And while losing a product (i.e., recycling or disposing of it at the end of its life) is often a smaller overall chunk of the emissions (around 6 percent of IKEA's, for instance), making sure a product is reused or recycled rather than thrown away has a huge impact in reducing waste overall.

The major caveats to this rule tend to be (interestingly and, perhaps, unfortunately) the things that end up being our biggest life and financial decisions. As we will cover later in this chapter, big-ticket purchases like our homes and cars combine a huge "Make It" impact with an even bigger "Use it" impact because we (ideally) use them for over a decade. Generally, **the more expensive and consequential decisions we make in our lives have a disproportionately higher impact on the environment**.

How "Reduce, Reuse, Recycle" Comes into Play

And now, back to why "Reduce, Reuse, Recycle" is so powerful as a hierarchy of decision-making: If the majority of the impact of most things comes from making them, then we can act more sustainability if we first *reduce* the number of things we purchase, then *reuse* the things we have purchased as long as possible, and finally *recycle* or repurpose where possible at each thing's end of life. For the major, long-lasting purchases in our lives like our homes and cars, we

---

20 "Product Gallery," Water Footprint Network, accessed July 2024, https://www.waterfootprint.org/resources/interactive-tools/product-gallery/.

21 "The Climate Footprint across the IKEA Value Chain," IKEA, accessed July 2024, https://www.ikea.com/global/en/our-business/people-planet/value-chain-climate-footprint/.

can use a data-driven approach to figure out how long we should use what we already have before switching to something more sustainable.

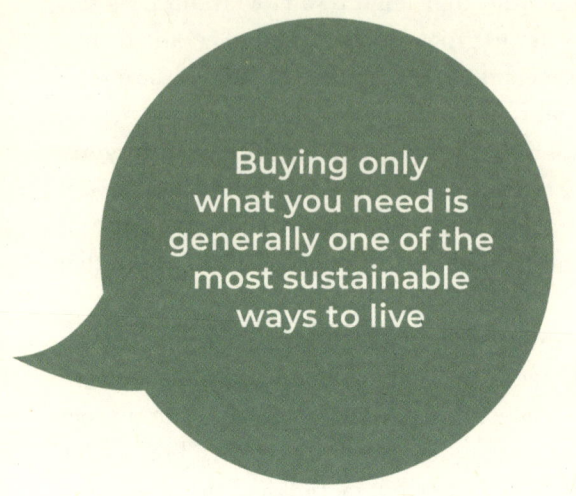

Buying only what you need is generally one of the most sustainable ways to live

Let's use a common example to illustrate this. While recycling a plastic water bottle is good, using a reusable water bottle is even better. The best scenario, however, would be to need as few water bottles as possible. To expand on that last point more generally, even if a product brands itself as being more sustainable, ultimately the most sustainable thing to do is to *reduce* and not buy it if you don't need it!

With this structure in mind, let us turn to a ranking of the top decisions we make in our personal lives and to what we can do to turn toward more sustainable activity. Some of these decisions are ones we make every day. Others we make once a year, and still others once in a lifetime. But all of them are valuable to consider the impact of.

Discussion Points

1) What do you think your "personal life emissions" pie chart may look like? What elements may be bigger or smaller for you in particular?

2) Where could you apply the hierarchy of "Reduce, Reuse, Recycle" in your life today? Are there places where you find yourself buying things you don't need?

3) What are some things in your life where you might not have considered the negative impact of making them (e.g., a car, a television)? Does this framing change your perspective on some of your possessions?

Chapter 2: Your Family

The single biggest lever we have for sustainability in our personal lives by a wide margin comes from our family planning. This may be shocking to some people until we remember that it is ultimately people who purchase homes, drive cars, take vacations, consume food, and create waste. So before we talk about the other decisions we make, it is prudent to start with the most impactful one.

According to the European Commission's "Emissions Database for Global Atmospheric Research," if you divide America's total emissions by its population, the average American incurs around 18 tons of greenhouse gas emissions annually.[22] And since an average American lives to be about seventy-seven years old,[23] you can roughly estimate that an American, with today's emissions per capita held constant, will emit around 1,400 metric tons of carbon dioxide and other greenhouse gases over their lifetime. This is roughly the same amount of carbon sequestered by 1,600 acres of forest in a year.

When we talk about families, we can't forget our furry friends, either. They are a part of this equation, albeit with a smaller impact. In a 2019 study, an average dog in the Netherlands was estimated to incur between 5 and 15 tons of carbon dioxide over their lifetime, or roughly the emissions from driving an SUV for two years.[24] Cats, being smaller (though they live longer lives), tend to incur fewer emissions, closer to 5 tons of carbon dioxide.

Hopefully, these numbers will decrease for future generations—in fact, that is looking increasingly likely and follows recent historical trends. Per the same data from the European Commission, America's per-capita emissions have dropped from 27 tons in 1979 down to 18 tons today. Our electric grids are decarbonizing at a faster clip, and major governments, cities, and corporations have committed to becoming net-zero within the lifetime of today's children.

---

22 "Country Fact Sheet: United States," EDGAR—Emissions Database for Global Atmospheric Research, accessed July 2024, https://edgar.jrc.ec.europa.eu/country_profile/USA.

23 Centers for Disease Control / National Center for Health Statistics, "Life Expectancy in the U.S. Dropped for the Second Year in a Row in 2021," August 31, 2022, https://www.cdc.gov/nchs/pressroom/nchs_press_releases/2022/20220831.htm.

24 Amélie Bottollier-Depois, "Carbon Pawprint: Is Man's Best Friend the Planet's Enemy?," Phys.org, March 20, 2021, https://phys.org/news/2021-03-carbon-pawprint-friend-planet-enemy.html.

An optimist can (and should) assume that the impact of the average human being and their pet(s) will shrink over the coming decades, so the "true" per-person impact will be significantly smaller. But it will be long (if ever) before our impact becomes zero.

To be clear, in the meantime, the wrong conclusion here is for all of us to stop having children. The societal problems that come from a drastically imbalanced demographic spread would be immense. (And it would invalidate the premise at the outset of the book to only consider solutions that work within our modern society.) Further, rearing a family is often influenced by our personal, cultural, and/or religious values, and those should not be casually cast aside. Instead, **the right conclusion is to be more intentional about family planning***.

Prior to proceeding, I want to acknowledge that many people face systemic barriers to family planning, especially those in communities that restrict access to sex education, lack availability of family-planning resources, or limit bodily autonomy (to name a few). I encourage the reader to seek additional resources in those cases and for all to keep these in mind for the policy section later in this book.

Here are three considerations for starting or growing your family in a more climate-friendly way:

1) **Adopt:** One of the single most climate-friendly things you can do is also one of the most benevolent. There are nearly half a million parentless children in foster systems and orphanages throughout America today. In addition to the incalculable impact you would have on one of these children, choosing to adopt means growing your family without needing to create new people. Every child you adopt versus every child you conceive is like preventing another 1,400 tons of carbon from entering the atmosphere.

   The same logic applies to our pets. Adopting a pet from a shelter not only carries a humane impact, but it comes with a lower carbon footprint than if you choose to breed your pet.

2) **Downsize:** The next biggest impact you can have in planning your family is to plan a smaller family. In addition to having fewer people, this gives you greater flexibility in making other more sustainable decisions, such as living in a smaller house or owning a smaller, more fuel-efficient vehicle. Plus, the fewer children you have, the smaller generational impact this can induce. Having only two children to then procreate themselves generally leads to fewer humans than having, say, five children who may each wish to start their own families.

3) **Delay:** One option that might not be fully intuitive here is having children later in life. In fact, over several generations, the impact of having three children by age thirty (and then those three children each having three of their own by age thirty and so on) is actually *smaller* than having two children by age twenty (and then those children having two of their own at age twenty and so on). The reason for this is that, when looking across generations, the rate of reproduction has a more dominating exponential growth effect than the amount of reproduction. Plus, if we believe that the world will become *more* sustainable over time, having kids later in life means they're born into a world with more renewable energy and more lower-impact products available to them.

For completeness, I should also mention the fourth option that many couples consider: deciding against starting a family in the first place. For some people, concerns about the planet's future may be so strong that the thought of bringing new life into an uncertain world is scary. I have several friends who have cited this as a reason for them and their partners to stay child-free, and while it's certainly not for everyone, it is a way to mitigate one of the biggest sources of emissions you can individually produce. And for those who still want some form of childlike companionship in their homes, the carbon footprint of having a pet is only 1 percent of that of a person over their lifetimes.

And in case it wasn't obvious: **Try to avoid fire-based gender-reveal parties if you decide to grow your family**. The El Dorado wildfire in California in 2020 was caused by a pyrotechnic device mishap at a gender-reveal party and ended up burning over twenty thousand acres of forest—and generating tons of global warming carbon emissions.[25]

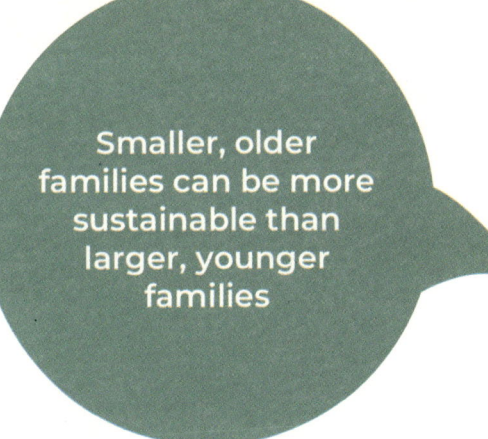

Smaller, older families can be more sustainable than larger, younger families

---

25 California Department of Forestry and Fire Protection, *2020 Wildfire Activity Statistics*, accessed July 2024, https://www.fire.ca.gov/incidents/2020/.

Discussion Points

1) How has climate change influenced the way you think about your family plans?

2) Do the details in this section change the way you think about what you may want for your family in the future?

Chapter 3: Your Home (39 Percent of Your "Personal" Emissions)

The next biggest decision most of us make in our lives comes in our decision of where to live and what type of home to live in. Generally, most families will only live in a handful of homes in their lifetimes, so the number of decision points on where you end up living tends to be fewer and further between. But for most of us, we are yet to live in the last location that we'll end up living in. For the purposes of this section, let's assume that you have the flexibility to live anywhere in the United States, though we know reality is often far more constrained.

I think that, intuitively, we have a sense of what a "more sustainable" house looks like at a general level. For example, a more modest home certainly feels more environmentally friendly than a McMansion. But there are many more factors—some obvious, others less so—that can help us in choosing a more sustainable home. You can boil this down to three* general tenets:

1) Location

2) Size

3) Efficiency

I am tacitly assuming that the reader is looking to buy an existing home rather than build a new one. If a reader is looking to build, then they would also need to consider the sustainability of the materials and of the architectural design for the new build.

Location

This is the most consequential of the three tenets, since a home's location will ultimately dictate some of the other types of impact that you will incur daily. There are decisions we can make here at both a higher level (e.g., where in the country should we live?) as well as specific (e.g., where in the city or town should I live?).

Is Your Home's Location Climate-Resilient and Climate-Friendly?

First, at the most basic level, it is worth understanding whether the home's location itself is climate-resilient and climate-friendly. For example, when it comes to climate resilience, is your home in an area that is water-rich or water-starved? Is it in a location that is relatively safe from natural disasters or rising water levels? In other words, is your neighborhood suited for sustainable habitation? Or is it at risk of climate impact?

Over the last decade, the American Southwest has been in a climate-change-fueled megadrought. Some of its most critical water resources that support habitation, such as Lake Mead in Nevada and Arizona, have been in the news about how much they have shrunk. Further (as we'll discuss in a little bit), homes in drier climates often use more water, which compounds the problem. Building a home or moving to a region that does not have a sustainable, consistent source of water puts pressure on a system that is already strained, threatening an eventual collapse of the water supply.

Likewise, in the American West (and now even the Upper Midwest and the Northeast), climate-change-fueled forest fires are decimating natural forest cover. Their increasing intensity and range are putting more communities at risk of fires and/or unhealthy air conditions. And along the Florida coast, homes are running out of insurers willing to insure them against flooding and storms as the intensity and frequency of major storms like hurricanes increase.

This is not to say that the people who already live there are wrong to live there. But if you are starting with a blank slate and looking where in America to live, it is now (sadly) prudent to ask whether the location being considered is climate-resilient. We have not yet seen the true first tidal wave of "climate refugees" within America, but when—not if—the time comes that certain places in the world stop being habitable by humans due to climate change, those humans will flee to safer climates. Their homes will invariably be left behind, unused and wasted.

Is Your Location's Infrastructure Climate-Friendly?

This is another consideration for where to live: Does the region where you live (or where you plan to move to) have available climate-friendly infrastructure? Or has the community there indicated a clear willingness to invest in this for the future?

One infrastructure consideration is the density of the region you are moving to and availability of public transport and other climate-friendly services. Cities like Chicago, New York, and Boston have robust rail and subway transit systems that help minimize the amount of reliance on cars. Other cities such as Houston, Phoenix, and Los Angeles are built to be more car-oriented, which creates far more reliance on higher carbon services and transit.

When it comes to the infrastructure behind the energy powering your home, different parts of the country power their electricity grids with varying levels of clean energy. States like Washington and New Hampshire leverage a lot of lower-carbon energy sources such as hydroelectric power, nuclear energy, wind, and solar to power their grids, whereas Wyoming, Kentucky, and others are much more reliant on coal and gas.[26] So prior to moving, it is worth checking whether the electricity provider in your region provides clean energy automatically or offers services to allow you to purchase renewable energy specifically. This information is often made available online.

Something to balance within this equation is that, on average, **heating your home takes more energy than cooling your home**. According to the EIA in its 2020 Residential Energy Consumption Survey, the average home energy consumption in cold climates (which require more heating) can be 62 percent higher than energy consumption in warm climates (which require more cooling).[27] Thankfully, there are many ways to reduce this gap, such as improved insulation and smart temperature management, both of which we'll get into later in this chapter.

Per this survey, the least energy-intensive region of America to live in tends to be the Pacific (with Hawaii being the "best" state), whereas the most energy-intensive regions are the Midwest and Alaska, the coldest state in America. In some regards, this makes sense. Heating your home to 70 degrees Fahrenheit when it is 20 degrees Fahrenheit outside requires enough energy to change the temperature by 50 degrees. Cooling, on the other hand, is usually trying to bridge a smaller gap (e.g., a 90-degree day only requires 20 degrees of cooling to get to that same 70-degree Fahrenheit indoor temperature).

---

26 "State Electricity Generation Fuel Shares," Nuclear Energy Institute, last modified August 2022, https://www.nei.org/resources/statistics/state-electricity-generation-fuel-shares.

27 "2020 RECS Survey Data," Residential Energy Consumption Survey (RECS), US Energy Information Administration, last modified March 2024, https://www.eia.gov/consumption/residential/data/2020/index.php?view=consumption#summary.

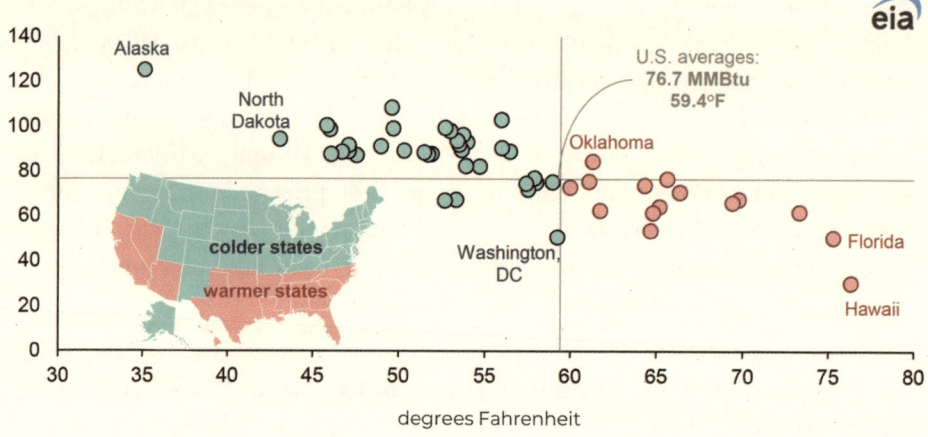

Source: "U.S. Households in Warmer States Consume Less Site Energy Than Households in Colder States," US Energy Information Administration, https://www.eia.gov/todayinenergy/detail.php?id=56380.

However—and this is why calculating true emissions impact is so darn tricky—just because houses in one region use more energy than houses in another *doesn't necessarily guarantee* that they create higher emissions. It is usually a pretty fair assumption to equate more energy to more emissions, but not all types of energy are created equal. When looking at all sources of energy globally, certain fuel sources for electricity (e.g., coal, oil) have far more deleterious effects than other types (e.g., nuclear or renewable energy). Even if you're using less energy, if that energy is really dirty, it can create more greenhouse gas emissions than using more (but cleaner) energy.

Using data from the EIA, if you take the energy-related carbon dioxide emissions (i.e., CO_2e) from the residential sector in each state and divide it by its population, you can get a "per household" view of average house emissions (not just energy usage) per state (see the appendix for the full list). Generally, colder states have higher residential carbon dioxide per capita than their warmer counterparts—not just because they require more energy to heat, but also because there often aren't "more sustainable" alternatives for using natural gas to heat their homes like there are with electricity, which can use solar or wind energy (and which is used in AC). The newest exception to this is electric heat pumps, an electricity-based heating solution subsidized by the passage of the Inflation Reduction Act of 2022.

That being said, you do occasionally find anomalies, such as how Oregon (with nearly half its energy from hydroelectric power) emits roughly the same per-household emissions as Georgia, how Washington, DC (with one of the country's highest percentages of solar usage) is more emissions-friendly than Oklahoma, or how South Dakota (with over half its energy from wind) is more emissions-friendly than Kansas, its warmer neighbor two states south. But generally, warmer states produce fewer household emissions on average.[28]

In short, if you have the option of where in the US you can live, **picking a location that is both more climate-resilient (i.e., safer from natural disasters or water scarcity) and climate-friendly (i.e., powered by cleaner electric grids and transit options and requiring less heating) tends to enable a baseline of lower carbon living**, all else being equal in your life.

Does Your Location Minimize Your Driving?

When you are choosing where in a city or town to live, the location of your home should also minimize your driving distance to work, shopping, and your hobbies. For an average vehicle with only one driver and no passengers, a single gallon of gas will emit around 9 kilograms of carbon dioxide into the atmosphere (the equivalent of charging a smartphone around one thousand times).[29] So any location you can choose that minimizes the distance to your most frequented locations (e.g., your work, the supermarket) will thereby reduce the emissions from your daily driving.

Let's say you're choosing between two homes: one that's in the city and takes 1 gallon to drive to work round-trip, and one that's in the suburbs that takes 2 gallons to drive round-trip. If you assume you're commuting five days a week (at the high end), then the extra emissions from living farther from your work could add up to 2 extra metric tons of greenhouse gas emissions from driving annually. This is the equivalent of 2.5 acres of forest.

Generally, unless your work and/or hobbies are far outside of a city, **you are almost always making the more sustainable choice by choosing to live within a city or a town rather than outside of it**. Not only does this cut down on a significant amount of driving emissions due to closer proximity, but it also potentially unlocks much more environmentally friendly transportation alternatives, whether that be trains, buses, cycling, or even walking. Further, city and town living tends to be less carbon-intensive due to higher population density, which allows more people to take advantage of the same

---

28 "Energy-Related CO2 Emission Data Tables," Environment, US Energy Information Administration, last modified July 12, 2023, https://www.eia.gov/environment/emissions/state/.

29 "Greenhouse Gas Emissions from a Typical Passenger Vehicle," United States Environmental Protection Agency, last modified on August 28, 2023, https://www.epa.gov/greenvehicles/greenhouse-gas-emissions-typical-passenger-vehicle.

shared infrastructure (e.g., roads, water pipes, telephone poles). Per research from the University of California, Berkeley, living in a city can be two to three times less emissions-intensive than living outside of one.[30]

The exception is if you need to spend a significant amount of time outside of a population center. If you work in the countryside, for example, chances are you're better off living there too. And if you're an avid outdoorsperson (e.g., skiing, climbing, hiking, biking, hunting), consider the irony that one of your biggest "get out into nature" activities may be one of the biggest emissions emitters if you're commuting from a city to do these activities versus living closer to the outdoors.

Does Your Location Minimize Your Long-Distance Travel?

Finally, the location of your home should minimize the amount of long-distance travel that you will need to do. Again, this applies more to decisions about where in the country makes sense to live rather than where in a particular city or town you should live. This ends up being a significant (and perhaps unconsidered) factor in our emissions.

I'll give a personal example. In 2020, I moved to Seattle for work. At the same time, most of my family and friends remained in the Midwest, where I am originally from. Between all the weddings and holidays that happened back in the Midwest in the years following, I must have flown back there a couple of dozen times. If an average flight from Seattle to Minneapolis generates about 150 kilograms of CO_2e per person, for example, that would translate to over 3 metric tons of carbon dioxide that would have been avoided had I still been living near my family and friends. (I have since moved back to the Midwest, which will help me on this front.)

This logic can apply to your work and holidays. If you have a job that expects to send you to a specific city with any frequency, is it possible to live closer to that city so that you can take other transit to get there? If you have a favorite vacation spot, can your home's location enable you to get there by train instead of by plane or car?

As I researched this segment, one data point surprised me: **At a sufficient distance, driving is not guaranteed to have a lower impact than flying**. A full flight from Chicago to Atlanta ends up generating about 100 kilograms of carbon dioxide per person. A solo drive in a car with 25 miles per gallon would end up burning around 28 gallons of gas, which—at a rate of 9 kilograms of carbon dioxide per gallon—ends up emitting 250 kilograms

---

30 Nadja Popovich, Mira Rojanasakul, and Brad Plumer, "The Climate Impact of Your Neighborhood, Mapped," *The New York Times* (online), December 13, 2022, https://www.nytimes.com/interactive/2022/12/13/climate/climate-footprint-map-neighborhood.html.

> Living closer to your work, errands, and hobbies is generally more sustainable than living farther away from them.

of carbon dioxide. That's over twice the amount as the flight! So unless you're traveling with at least three people in your car, or if you have a car with good gas mileage or that runs on electric power, flights can actually be *more* sustainable than driving because of how many more people they can move at once versus a mostly empty car. We will dive more deeply into the ins and outs of emissions from our transportation later in this chapter.

For this reason, the most impactful thing you can do to minimize the emissions of your long-distance travel is to minimize the distance you need to travel, either via how and where you choose to live, or on where (and if!) you choose to travel.

Discussion Points

1) Do you feel like you live in a climate-resilient location today? What are some types of climate-driven stressors that could impact where you live in the future?

2) When you're away from home, where do you spend most of your time? What are some things you could change to help minimize the amount of distance you need to travel?

3) What options do you have to spread out your emissions? Can you carpool? Take a shuttle? Use on-demand public transportation that's available to you?

Size

The next biggest indicator of emissions for your home tends to be the size and form of the house (and yard!) itself.

How Your Home's Size Contributes to Emissions

Based on analysis from the EPA, the average home emits around 8 metric tons of carbon dioxide annually (nearly 10 acres worth of forests!) from heating, cooling, and using electricity for appliances, lights, televisions, and the like.[31] About 5 of those tons come from the house's electricity use (which includes AC), and the remainder comes from heating the house and its water using other sources like natural gas. This number will fluctuate depending on the climate you live in, with homes in colder climates generally inducing more emissions from heating.

As you can probably guess, the bigger the house, the more energy you need to maintain it, whether to fully heat the home, cool the home, or keep all the lights on. To give a sense, home electricity usage is often measured in kilowatt-hours (kWh). You may be familiar with wattage from when you have replaced a light bulb or plugged in an electronic; it's the maximum amount of power the device can output. Kilowatt-hours, then, represent the amount of energy it takes to run that device for one hour. If you have a 100-watt lightbulb turned on for ten hours, it will consume 1 kWh of energy. (A kilowatt is 1,000 watts.) Based on a survey of some of the appliances in my home, 1 kWh of energy roughly equals a day's use of a fan, nine hours of watching TV, an hour of using my microwave, and eight hours of using my refrigerator. Emissions-wise, every kilowatt-hour you use is about equivalent to driving your gas vehicle 1 mile*, so the lower the wattage for a device, the better.

It is this last point that makes it possible for an electric car to power your entire home for over a day, as you may have seen in ads for Ford's first electric truck, the F-150 Lightning. The relative energy your car uses compared to your entire house is surprisingly high.

Heat works a little differently. It is not always electricity-based, so it isn't generally measured in kilowatt-hours. About half of households burn gas directly to heat the home.[32] You may also be using gas for your stovetop or heating your water with gas. As we discussed earlier, these types of emissions are worse, since **heating is often worse for the environment than cooling**.[33] Per the EPA, the average home in America emits around 2.5 metric tons of carbon dioxide annually from heating—roughly the equivalent of cutting down

---

31 "Greenhouse Gases Equivalencies Calculator—Calculations and References, Home Energy Use," United States Environmental Protection Agency, last modified May 17, 2024, https://www.epa.gov/energy/greenhouse-gases-equivalencies-calculator-calculations-and-references#houseenergy.

32 Kaili Diamond and Matthew Sanders, "The Majority of U.S. Households Used Natural Gas in 2020," US Energy Information Administration, March 23, 2023, https://www.eia.gov/todayinenergy/detail.php?id=55940.

33 US Energy Information Administration, "Table CE1.1 Summary Annual Household Site Consumption and Expenditures in the United States—Totals and Intensities, 2020," last modified March 2024, https://www.eia.gov/consumption/residential/data/2020/c&e/pdf/ce1.1.pdf.

forty trees a year per household (split between heating the house and heating water).[34]

As for why this is—and in addition to the fact that heating is often trying to cover for a larger temperature differential—consider that you can use any type of electricity (including renewable energy!) to cool your home, but many of us can only use gas or other fuels to heat our home. Most grids have at least *some* renewable or lower-carbon energy powering them, so you tend to save on emissions by using electricity instead of fuel. This logic also applies to why electric vehicles (EVs) tend to have lower emissions than gas-powered vehicles. It is only in extreme cases where an electric grid is almost exclusively powered by energy like coal—which is "dirtier" than gas in terms of emissions—that electric emissions can be more intense than heating from natural gas.

Thinking in terms of these two energy types, you can start to see how emissions roughly scale with size, as a bigger house needs more electricity for air conditioners, light bulbs, and vacuum cleaning (if tidy!) as well as more gas to heat the larger space. Per the EIA, the average household between 2,500 and 3,000 square feet uses roughly 30 percent more energy than a house that is between 1,500 and 2,000 square feet.[35] Thankfully, a house that is twice as big doesn't need twice the amount of energy to run. Nonetheless, a smaller home will be more carbon-efficient.

How the Number of People Living in Your Home Contributes to Emissions

The one consideration that can invert this logic is the utilization of the home—specifically, how many people live in it. The more people living in a home, the better your per-person emissions from said home become. It is almost always more sustainable to live with others than to live alone. In this case, if you're in a 2,500- to 3,000-square-foot home because you have a bigger family living together (say, five people), the per-person emissions of the home can actually be smaller than a 1,500- to 2,000-square-foot home occupied by only a few people. This logic is why, on average, condos and apartment complexes are often the most sustainable types of housing. You can fit way more people into their overall square footage than in any other type of housing and take advantage of way more shared infrastructure.

Here's another reason why utilization is important: **Beyond the annual energy needed to maintain your home, a considerable amount of**

---

34 US Energy Information Administration, "Table CE4.1 Annual Household Site End-Use Consumption by Fuel in the United States—Totals, 2020," last modified March 2024, https://www.eia.gov/consumption/residential/data/2020/c&e/pdf/ce4.1.pdf.

35 US Energy Information Administration, "Table CE1.1."

emissions was required to build your home in the first place. While there isn't *great* data available on this (the EPA's most up-to-date analysis on this relies on data from 1998), a group called the Rocky Mountain Institute, a reputable sustainability research nonprofit, estimated in 2023 that an average home in America contains around 50 metric tons of embodied carbon, roughly the same amount of emissions it takes 60 acres of forest to absorb in a year.[36] Embodied carbon includes everything from the emissions of manufacturing and transporting the building materials, to the construction of the home, to the maintenance of the household, to its eventual demolition at end of life. It excludes, however, the energy needed to heat, cool, and use electricity (these things are not "embodied").

Per our earlier logic, this number goes up as the house gets bigger. Building a new home generates the same emissions as living in it for over six(!) years, so anything you can do to spread out those sunk building costs across multiple people becomes beneficial. If you have three people living in a home, the per-person building emissions become six divided by three, or 2 metric tons of CO_2e per person. (Or, better yet, buy an existing home instead of tearing a perfectly functional one down to build a new one!)

> Maximizing the number of people per square foot in a home is a more important impact lever to improve sustainability than the size of the home itself.

To generalize the guidance here, **whatever you can do to minimize your "square footage per person" in your home will likely reduce the amount of emissions per person that your home produces**.

How Your Home's Lawn Contributes to Emissions

The other "size of house" consideration is lawns, which is the biggest irrigated "crop" in America.[37] Lawns are a bit of a different beast, as they don't themselves require heating or cooling, nor are they plugged in. However, most people with lawns are likely making decisions about how often to water these lawns, how often to mow them, and the kind of biodiversity

---

36 Chris Magwood, Tracy Huynh, and Victor Olgyay, "The Hidden Climate Impact of Residential Construction," RMI (Rocky Mountain Institute), accessed July 2024, https://rmi.org/insight/hidden-climate-impact-of-residential-construction.

37 Cristina Milesi, Steven Running, Christopher Elvidge, John Dietz, Benjamin Tuttle, and Ramakrishna Nemai, "Mapping and Modeling the Biogeochemical Cycling of Turf Grasses in the United States," Environmental Management, 2005, https://link.springer.com/article/10.1007/s00267-004-0316-2)

that's present. Depending on these factors, a bigger yard often generates more negative impacts on the environment. And while a lawn *can* be a carbon sink (i.e., removing more emissions from the atmosphere than they use), they are often not because of the amount of negative impact associated with their upkeep, not to mention their impact on local biodiversity and potential issues with runoff and erosion.

Unfortunately, there are no reliable, up-to-date sources of data on the emissions from lawn upkeep in the United States (at least not that I could find). The most recent dedicated publication from the EPA is a 2015 study, which estimated that 26.7 million tons of pollutants were emitted by gas-powered lawn and garden equipment in 2011.[38] Though the carbon dioxide portion of these pollutants was estimated as only 4 percent of all carbon dioxide emissions in America that year, because lawn mower engines were far less regulated (and therefore less efficient) at that time than car engines, a lot of those pollutants were far dirtier than what a car would produce. Our lawn care equipment in 2011 produced nearly 17 percent of what are known as "volatile organic compounds" in America annually. This is far worse than carbon dioxide for both the environment and our own health when we're exposed to those compounds.

While lawn equipment built today will likely be more efficient than what we saw a decade ago, it is fair to assume that the impact of our lawn tools is, at minimum, what you would expect from a car using the same amount of gasoline. A gallon of gas used in tending your lawn will at least have the same impact as driving an average car 25 miles. If you have all-electric equipment, you can mitigate most of this impact if your home is powered by cleaner energy to charge your equipment. (Manual equipment that doesn't require fuel or energy, of course, would be best.)

While data on lawn care emissions is scarce, we thankfully have a little more detail (but not a lot more) on lawn care water usage. Per a 2016 study by the Water Research Foundation[39] and estimates from the EPA,[40] we can estimate that an average home in America uses between 30 and 50 percent of its total household water usage outdoors, with this number varying wildly depending on the local climate. Using water as a focus in this case, rather than emissions, can help us understand the most sustainable ways to tend to our lawns and gardens.

---

38 Jamie L. Banks and Robert McConnell, "National Emissions from Lawn and Garden Equipment," United States Environmental Protection Agency, 2015, https://www.epa.gov/sites/default/files/2015-09/documents/banks.pdf.

39 Water Research Foundation, "Residential End Uses of Water, Version 2", April 2016, https://www.circleofblue.org/wp-content/uploads/2016/04/WRF_REU2016.pdf.

40 "How We Use Water," United States Environmental Protection Agency, last modified April 3, 2024, https://www.epa.gov/watersense/how-we-use-water.

For those who don't know the history of lawns in America, it's quite an interesting story as to how we became so lawn-obsessed. According to an interview with John Fleck, the director of the Water Resources Program at the University of New Mexico, our obsession with well-manicured lawns is a holdover from colonial times.[41] In seventeenth-century England, lawns became a symbol of status and wealth because of the high cost to maintain them, and English settlers who came to America brought those cultural values with them. It is possible that you live in a jurisdiction today that has codified this culture into laws specifically mandating a certain type of lawn. For example, in Lakeville, Minnesota (a suburb of Minneapolis), a yard shall be declared a "public nuisance" and a misdemeanor levied if grass or weeds exceed 8 inches tall in a yard.[42] Sadly, this particular focus on weeds (literally "any plant that is considered undesirable, unattractive, or troublesome") can often make illegal some of the plants that are most naturally suited to grow in an environment in favor of the homogeneity of grass lawns.

While grass lawns are suited for the climate on the East Coast (due to its high amount of consistent rain) where settlers first arrived, once Americans started moving west, they brought their lawns with them irrespective of whether the new climate could support them. In a vicious cycle, these western climates are largely unsuited to support grass lawns, making the amount of water needed to keep them alive far more taxing. According to a 2015 study from the USGS, total household water usage (indoor plus outdoor) ranges from pretty modest for colder, rainier states better suited for grass lawns (between 35 and 55 gallons per day for Connecticut, Vermont, Maine, and Wisconsin) to pretty severe for warmer, drier states that don't naturally support grass lawns (more than 145 gallons a day for Arizona, Utah, Wyoming, and Idaho).[43]

The great irony is that the more an area is starved for water, the more its lawns require a larger amount of the already limited supply. The same general principle applies to golf courses. While a golf course in the desert might be a beautiful place to play, it is, ecologically speaking, very problematic in its water usage.*

*Even for places that may seem water-rich today, climate change runs the risk of more frequent droughts in the future, making water conservation relevant to all of us. Further, even water-rich locations may be draining groundwater reserves at unsustainable rates. Humanity

---

41 Rachel Ramirez, "Why the Great American Lawn Is Terrible for the West's Water Crisis," CNN.com, April 29, 2022, https://www.cnn.com/2022/04/28/us/why-grass-lawns-are-bad-for-drought-water-crisis-climate/index.html.

42 "4-1-3: LENGTH OF WEEDS AND GRASS," n.d., American Legal Publishing, https://codelibrary.amlegal.com/codes/lakevillemn/latest/lakeville_mn/0-0-0-2033.

43 United States Geological Survey's Water Availability and Use Science Program, *Estimated Use of Water in the United States in 2015*, 2018, https://pubs.usgs.gov/circ/1441/circ1441.pdf.

has withdrawn so much groundwater from aquifers over the last few decades that Earth's axis tilt has measurably shifted over that time span.[44]

So what do we do about our lawns to reduce the amount of water and emissions they induce? It's best to turn to nature to guide our path.

Before humanity came in and imposed lawns on the local landscape, there was already a rich, natural ecosystem thriving in its place. If you can "give lawns back to nature" by reintroducing more native plant species, you can cut down dramatically on mowing emissions (as these plants likely don't require mowing) as well as water usage, assuming these plants are more tailored to the local climate and water availability. Further, more native plants will have far deeper root systems and offer more biodiversity, meaning that you'll likely get a more exciting swath of wildlife visiting your lawns while bolstering the strength of your soil. And while some jurisdictions 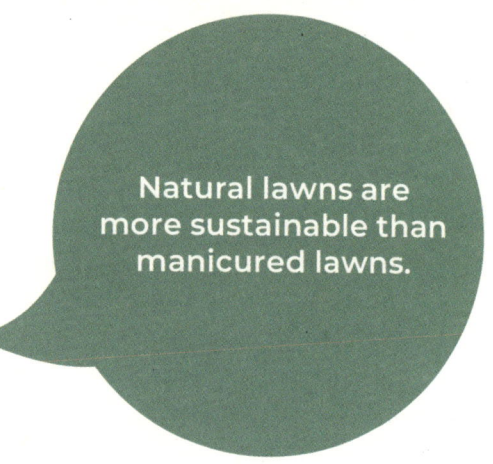 make this more difficult for you legally (e.g., Lakeville, Minnesota), others will actually pay you to make this change. San Diego, for example, will pay residents between $2 and $4 per square foot of yard that's converted from grass lawn to more natural landscaping.[45] In this case, it's a win for water, a win for emissions, a win for biodiversity, and a win for your wallet, all in one!

---

44 Ki-Weon Seo, Dongryeol Ryu, Jooyoung Eom, Taewhan Jeon, Jae-Seung Kim, Kookhyoun Youm, Jianli Chen, and Clark R. Wilson, "Drift of Earth's Pole Confirms Groundwater Depletion as a Significant Contributor to Global Sea Level Rise 1993–2010," *Geophysical Research Letters* 50, no. 12 (June 2023), https://doi.org/10.1029/2023GL103509.

45 "Turf Replacement," County of San Diego Waterscape Rebate Program, San Diego Department of Public Works, accessed July 2024, https://www.sandiegocounty.gov/content/sdc/dpw/watersheds/RebatesIncentives/Turf_Replacement.html.

> **Discussion Points**
>
> 1) How does the concept of maximizing "people per square foot" in housing influence how you think about your living arrangements? What kind of local policies could help encourage more sustainable living?
>
> 2) What are some ways in which you could reconsider how you treat your lawn, if you have one? Are there interesting alternatives for grass that you could feature instead?

Efficiency

Finally, after talking about the location and the size of the house, the last portion is the efficiency of the house itself. In this context, we'll define *efficiency* as "emissions intensity" as well as water intensity. There are a few ways to tackle this, with all of them having the bonus of helping reduce your energy bills. And the lower your energy bills get, the better you'll make things for the planet.

Reducing the Overall Amount of Energy and Water Needed to Run a Home

This first strategy largely centers on whether the energy used to heat or cool that home is being optimally used. In an extreme example, heating your home to 70 degrees Fahrenheit in the winter while having all your windows open will require you to burn more energy than if they were closed because a lot of the heat being produced escapes. However, there is *a lot* we can do before we even start to talk about buying new, more efficient heating, ventilation, and AC systems, which we'll talk about later.

Even if our windows are closed, houses are large, complex structures—and, especially for older homes, there are likely gaps in how well they are insulated. The activity of getting your home ready to reduce your heat needs (and, therefore, your energy bills) is called "winterizing" your home. The US Department of Energy (DOE) helpfully lists a number of ways to make sure that the heat you're producing in the winter actually stays within your home[46] (thereby requiring less of it), such as:

---

[46] "Fall and Winter Energy-Saving Tips," US Department of Energy, accessed July 2024, https://www.energy.gov/energysaver/fall-and-winter-energy-saving-tips.

WHAT WE CAN DO

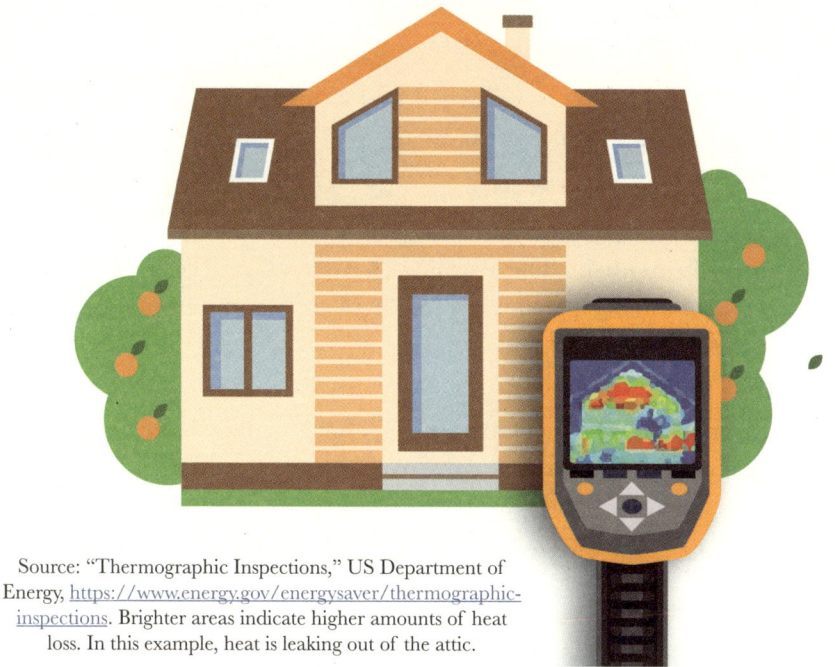

Source: "Thermographic Inspections," US Department of Energy, https://www.energy.gov/energysaver/thermographic-inspections. Brighter areas indicate higher amounts of heat loss. In this example, heat is leaking out of the attic.

- Opening curtains on windows when the sun is shining on them and closing them otherwise.

- Installing sheets or insulating drapes on windows to insulate them during the winter.

- Turning down the temperature on the thermostat when you are asleep or out of the house for an extended period. Google estimates that its smart thermostat, Google Nest, can save 10 percent on heating bills annually by actively managing heat in this way.[47]

- Keeping the thermostat lower. For every degree cooler you keep your thermostat over eight hours, the DOE estimates you can reduce your heating bill by 1 percent.[48] Keeping it lower over a full twenty-four-hour period could extend savings to 2 to 3 percent per degree.

- Finding and sealing leaks around utility cut-throughs for pipes; gaps around chimneys, doors, and windows; and unfinished spaces such as attics. Also, make sure to insulate your air ducts (which, if sealed im-

---

[47] "How the Nest Thermostat Savings Calculator Works," Google Nest Help, Google, accessed July 2024, https://support.google.com/googlenest/answer/9241995.

[48] "Programmable Thermostats," US Department of Energy, accessed July 2024, https://www.energy.gov/energysaver/programmable-thermostats.

properly, can leak up to 60 percent of heated air!) and clean to remove dust buildup.[49]

- Keeping the chimney plugged and closed if you own a fireplace but never use it. And if you do use it, keep the damper closed when it's not in use.

- Lowering the temperature on your water heater to "warm" (120 degrees Fahrenheit). The Washington Utilities and Transportation Commission estimates that every 10-degree reduction in water temperature can save between 3 and 5 percent in energy costs.

Not listed on the DOE's list is an idea that I tend to enjoy myself: **doing more oven-based cooking in the cold months**. Not only do you get to enjoy a hot meal during the cold season, but you can also leave the oven open after you're done to let the excess heat warm the room (and reduce the amount of work your boiler has to do). It's a great way to "recycle" heat and keep it from being wasted!

Another consideration is to use electric space heating if you don't need a lot of heat rather than spending energy to heat your entire home. Space heaters are a way to focus the heating just to where you are spending time in the home. You can also use a combination of heating sources, like keeping the thermostat at 50 degrees Fahrenheit (to keep pipes from freezing) and space-heating the room you are in.

> Every degree cooler you keep your home in the winter can reduce your heating emissions by 3 percent.

The DOE also has a similar set of recommendations for the warmer months to keep homes cool,[50] such as:

- Closing curtains when the sun is bearing down on them.

- Planting deciduous trees to help block some of the sunlight onto the home during the summer months, when leaves are still on the trees.

- Turning off the AC when you are out of the house for an extended peri-

---

49 "Why Energy Efficiency Matters," US Department of Energy, accessed July 2024, https://www.energy.gov/energysaver/why-energy-efficiency-matters.

50 "Spring and Summer Energy-Saving Tips," US Department of Energy, July 2024, https://www.energy.gov/energysaver/spring-and-summer-energy-saving-tips.

od. Similar to the above, a smart thermostat can save around 15 percent on cooling bills based on this behavior.[51]

- Keeping the thermostat higher in general, which is the inverse of winter. For every degree warmer you keep your thermostat over eight hours, you can reduce the AC portion of your electric bill by 1 percent.[52] Doing so over a twenty-four-hour period can yield 2 to 3 percent per degree.

- Cleaning your air intake vents and air ducts regularly to remove dust buildup.

- Keeping your home well insulated, just like in winter, to prevent cold air from leaking out and hot air from leaking in.

Other strategies around smart window and fan usage can also help. In the evening, if the temperature outside has dropped to 72 degrees Fahrenheit or lower, simply opening the windows can be a zero-energy way to cool the home. And, rather than feeling a need to cool the entire home, having a portable fan (or even a window unit for the rooms where you spend the most time) can be a great, more energy-efficient way to cool yourself off, just like a space heater can be to keep you warm in the winter.

From an emissions point of view, how much does lower matter? It turns out that it matters a lot! Applying all of the above suggestions could yield upward of a 15 to 20 percent reduction in your energy bills, per estimates from the DOE[53] and the EPA.[54] And that's before considering tax rebates available for doing some of the above (if still available) thanks to the Inflation Reduction Act of 2022.

Also, because energy bills are correlated to the amount of energy you use, you can roughly (though not perfectly) estimate a reduction in your home energy emissions by 15 to 20 percent. For an average home,

> More efficient heating and cooling strategies can reduce your home energy consumption by 20 percent or more.

---

51 "How the Nest Thermostat Savings Calculator Works," Google Nest Help, Google.

52 "Programmable Thermostats," US Department of Energy.

53 "Programmable Thermostats," US Department of Energy.

54 "Methodology for Estimated Energy Savings," Energy Star, United States Environmental Protection Agency, accessed July 2024, https://www.energystar.gov/saveathome/seal_insulate/methodology.

a 20 percent reduction across heating and cooling can reduce annual household emissions by around 1 ton![55] That's nearly the same as growing eighteen trees for a whole decade. If you're part of an average household in America, that means you'd be reducing your personal cut of emissions by between 400 and 500 kilograms (or seven trees) of greenhouse gases annually.

Reducing Your Home's Reliance on Fossil Fuels and Public Water

Unless you live in the Garden of Eden, at some point your home will have to use energy no matter how efficient you've gotten it. Thankfully, we now live in a society where we can mitigate the emissions of most of the energy needed to run our households. Let's start with the obvious: getting away from using fossil fuels to power our homes.

Some of us may be lucky enough to have our power come from a utility provider that lets us buy 100 percent clean energy. These programs, often called utility green power programs or green pricing programs, require no change on behalf of the consumer. Instead, the consumer would agree to pay a little bit extra for a guarantee that their home would either be directly powered by green energy or finance a renewable energy credit to generate the equivalent amount of renewable energy somewhere else. Per analysis from the EPA in 2015, the "green premium" you would have to be willing to pay comes out typically to around $18 per month.[56] (This would be less if you are successful in applying the energy-efficient techniques we've already talked about!).

For those of us who don't have access to this type of product, the other primary option (beyond exploring complex energy credit financial instruments) is to consider installing on-site renewable energy, typically solar.

With the passage of the Inflation Reduction Act of 2022, the cost of installing your own source of clean energy was made a little bit cheaper. Through 2032, if the Inflation Reduction Act remains in effect, all households are eligible to get a 30 percent tax rebate for the cost spent on solar panels as well as labor, permitting, inspection, and development costs.[57] With the current average cost of installing residential solar panels hovering around $20,000,[58] this can mean a discount of $6,000 (nice!), with some states offering additional in-

---

55 "Use of Energy Explained," US Energy Information Administration.

56 "Green Power Pricing," United States Environmental Protection Agency, last modified February 9, 2024, https://www.epa.gov/green-power-markets/green-power-pricing.

57 Courtney Lindwall, "A Consumer Guide to the Inflation Reduction Act," National Resources Defense Council, July 20, 2023, https://www.nrdc.org/stories/consumer-guide-inflation-reduction-act.

58 Brian Reil, Nathanael Greene, and Sean Gallagher, "Joint Statement of the Edison Electric Institute, the Natural Resources Defense Council and the Solar Energy Industries Association on Solar Power's Benefits and Prospects," Edison Electric Institute, February 15, 2022, https://www.eei.org/-/media/Project/EEI/Documents/Resources-and-Media/Newsroom/2152022EEI-NRDC-SEIA-joint-solar-statement-FINAL.pdf.

centives on top of that. That being said, $14,000 is still a pretty hefty up-front price to go fully renewable.

Thankfully, once installed, solar panels can eliminate most of your energy bill. In some cases, it can *make* you money if your grid has programs to purchase your excess energy from you! Taking data from the EIA, which shows the average annual household electric bill is $1,380.00, this means that the average solar panel will pay itself off in around ten years[59]. Given that the DOE estimates that most solar panels have a total lifespan of around thirty years, going solar can save you nearly $3,000 over the lifetime of the panels![60]

And if you own your home, that's not even accounting for a property value increase tied to the installation of the panels. A study from the home-buying platform Zillow found that solar panels increased the sale value of an average home by 4.1 percent (or over $9,000!).[61]

If an up-front cost of $14,000 is too hefty for you to switch to green energy, there are some financing options available that let you have solar panels installed free of charge. These arrangements are typically known as **power purchase agreements (PPAs)**. The terms of PPAs can vary, but almost all of these agreements will give you a cheaper electricity bill (with some solar companies claiming an average savings of 10 to 20 percent).[62]

While the longer-term return on investment (ROI) isn't as strong as purchasing your own solar panels, PPAs offer a free-of-cost way to power your home with renewable energy while reducing your electric bills. In other words, **anyone with a home can afford to go solar in some way, shape, or form**.

Fully powering your home with renewable energy can reduce the emissions of an average home by over 60 percent(!), or around 5 metric tons of greenhouse emissions annually. That's equivalent to the amount of carbon emissions sequestered by 6 acres of forest annually. However, the average American household emits 8 tons of greenhouse gas emissions annually. So why won't converting to green electricity reduce household energy emissions down to zero?

As we talked about earlier, renewable-generated electricity only works for things that actually use electricity. For most homes, the single greatest source of emissions tends to come from heating (first the home, followed by its water),

---

59 Peter Wong and Paul McArdle, "Average Monthly Electricity Bill for U.S. Residential Customers Declined in 2019," US Energy Information Administration, December 15, 2020, https://www.eia.gov/todayinenergy/detail.php?id=46276.

60 "End-of-Life Management for Solar Photovoltaics," US Department of Energy, accessed July 2024, https://www.energy.gov/eere/solar/end-life-management-solar-photovoltaics.

61 Sarah Mikhitarian, "Homes with Solar Panels Sell for 4.1% More," Zillow, April 16, 2019, https://www.zillow.com/research/solar-panels-house-sell-more-23798/.

62 Kerry Thoubboron, "Power Purchase Agreements: What You Need to Know," EnergySage, February 12, 2021, https://www.energysage.com/solar/power-purchase-agreements-overview/.

and that is achieved by burning natural gas in about half of US households per data from the EIA. Thankfully, there are options to switch your heating over to electricity, which then means it can be powered by renewable energy. And even if it isn't, the DOE estimates that heat pumps can be twice as energy-efficient as other forms of heating.[63]

As a bonus, heat pumps don't just pump heat *into* your house in the cold months. They also pump heat *out* of your house in the warm months, serving the dual role of heater and air conditioner all at once.

As for heating water, there are also heat pump water heaters available for purchase to replace traditional water heaters. Whether you are replacing a gas-powered water heater or an electric resistance heater, heat pump water heaters tend to be far more energy efficient—up to 300 percent![64] The resulting cost benefit can translate into a reduced energy bill of $550 a year, per estimates from the Natural Resources Defense Council using Energy Star data.[65] Given an up-front cost of just over $1,100, this means you could see a full payback in just a few years.

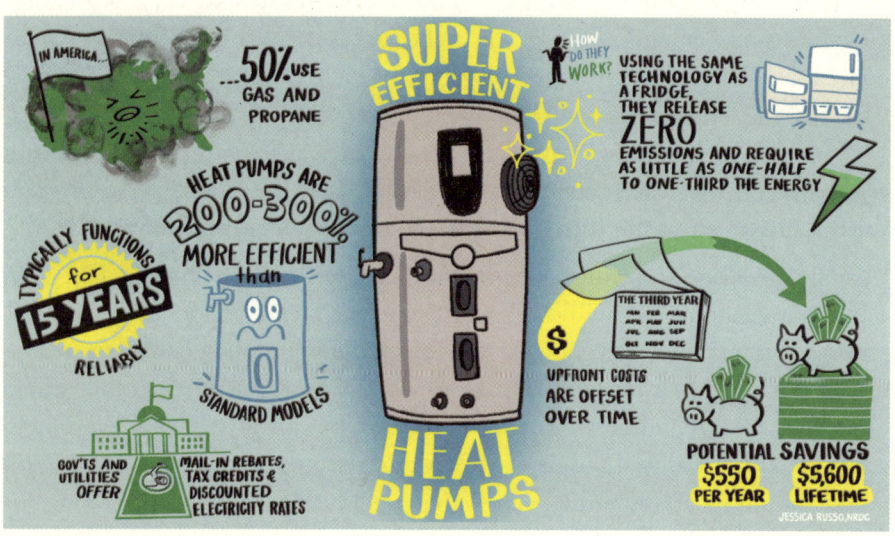

Source: "What's the Most Energy-Efficient Water Heater?," National Resources Defense Council, https://www.nrdc.org/stories/whats-most-energy-efficient-water-heater.

---

[63] "Heat Pump Systems," US Department of Energy, accessed July 2024, https://www.energy.gov/energysaver/heat-pump-systems.

[64] Patrick Rogers, "What's the Most Energy-Efficient Water Heater?," National Resources Defense Council, August 22, 2023, https://www.nrdc.org/stories/whats-most-energy-efficient-water-heater.

[65] "Save More with ENERGY STAR Certified Heat Pump Water Heaters," Energy Star, United States Environmental Protection Agency, accessed July 2024, https://www.energystar.gov/products/heat_pump_water_heaters/benefits-savings.

Thanks to the oft-mentioned Inflation Reduction Act (assuming it is not rescinded or amended), you can claim 30 percent of the cost of both types of heat pumps as tax rebates, capped at $2,000 annually through 2032 at the time of writing. There are additional rebates available for certain buyers of heat pumps, such as up to $8,000 for low- to middle-income families via the High-Efficiency Electric Home Rebate Act.[66]

With all these things combined, you can both almost fully electrify your home AND power that electricity with renewable energy, with far less cost than is often feared and far more savings than many people realize. **Altogether, this can result in annual household energy emissions reductions of almost 100 percent, or nearly 8 tons of emissions annually.** This is the equivalent of the carbon dioxide absorbed by nearly 10 acres of forest.

It's worth noting that, just like with everything else, there *is* some very real embodied carbon associated with the manufacturing of solar panels and heat pumps. Similar to what we will see with EVs in the next section, industry consensus from solar panel manufacturers is that it takes about three years for a solar panel to fully "offset" the emissions associated with its manufacturing.[67] And per analysis from the University of Liverpool, the payback period for heat pumps could be even faster![68] Generally, though, if you have a boiler that's close to the end of its life, it doesn't hurt to let it run its course before installing your new, greener alternatives.

> Shifting to renewable energy and electric heat pumps can nearly eliminate your home's energy emissions.

Before shifting gears (a poor pun in anticipation of our next section), let's close with how to reduce our reliance on public water. While there aren't nearly as many options for how to control the flow of water into your home as there are for electricity, there are, nevertheless, some creative ways to reduce your reliance on public water, which is not always sustainably drawn.

---

66 Lindwall, "A Consumer Guide."

67 Sam Wigness, "What Is the Carbon Footprint of Solar Panels?," Solar.com, August 31, 2023, https://www.solar.com/learn/what-is-the-carbon-footprint-of-solar-panels/.

68 Stephen Finnegan, Craig Jones, and Steve Sharples, "The Embodied CO2e of Sustainable Energy Technologies Used in Buildings: A Review Article," *Energy and Buildings* 181 (September 2018): 50–61, https://doi.org/10.1016/j.enbuild.2018.09.037.

Outside of more obvious ways to reduce water usage like taking shorter showers (we'll dive more deeply into this and other points later), your main source of opportunity relies on rainwater. If you live in a place that experiences a decent amount of rainfall, consider setting up outdoor rain barrels or cisterns, if allowed. Once that water has been filtered, it can be used to replace public water usage for many non-potable uses, from watering your yard and garden on dry days to washing your car, flushing your toilet, or washing your clothes. And if you are flush with cash, there are some incredibly cool (and incredibly expensive) "gray water recycling" systems that can cover nearly all types of water usage.

Discussion Points

1) Have you ever tried winterizing or "summerizing" your home before? What are some tricks and practical steps you can take to help retain your heating and cooling?

2) How do you see the relationship between energy efficiency in your home and renewable energy usage to power your home? Does doing one mean you don't need to do the other?

Chapter 4: Your Travel and Transportation (33 Percent of Your "Personal" Emissions)

Your Vehicle and Transportation Patterns

Next up on the impact list after your family and your home comes your travel and transportation. Per the EPA, **greenhouse gas emissions from transportation (including shipping of goods) account for about 29 percent of total US greenhouse gas emissions**. This makes it nearly tied as the largest contributor of US greenhouse gas emissions with industry (30 percent) and commercial and residential emissions (31 percent).*

*Normally, industry would be the largest contributor to emissions in a country. But over the past century, America has outsourced the dirtiest parts of its supply chain to other countries. Thankfully, the Greenhouse Gas Protocol is working to better allocate emissions to the countries responsible for their emissions no matter where they take place.

Within transportation, 57 percent of emissions come from passenger vehicles (with around twice as much coming from SUVs and trucks than sedans or other cars), and just under 2 percent comes from passenger air travel. The fact that traveling by plane is such a small portion of America's emissions compared to driving might stand out as counterintuitive, but keep the following in mind:

1) There are *far* fewer people who fly versus drive.

2) There are *far* fewer planes than cars.

3) As mentioned earlier, flying can be more carbon-friendly than driving for certain long distances due to a plane's capability to transport many people at once.

U.S. EMISSIONS PER TYPE OF TRANSPORTATION (2022)

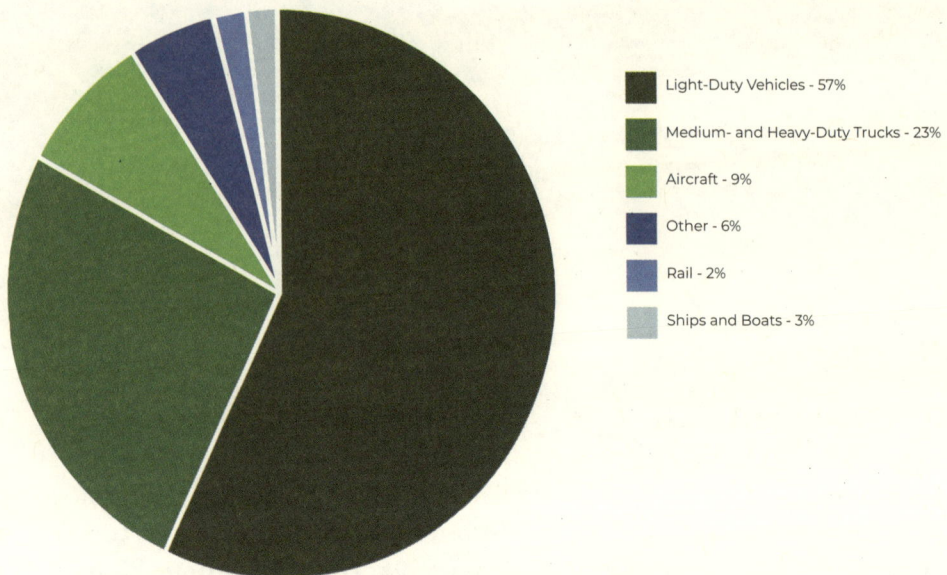

Source: "Fast Facts on Transportation Greenhouse Gas Emissions," US EPA, accessed July 2024, https://www.epa.gov/greenvehicles/fast-facts-transportation-greenhouse-gas-emissions.

For some of us, our emissions from transportation and driving actually exceed the emissions that our homes produce! Though the EPA indicates that the average vehicle miles traveled in 2019 was 11,520 miles per year,[69] if your household drives more than 20,000 miles a year in a 23 miles-per-gallon car, then your car likely emits more greenhouse gases annually than your house. Or, if you travel for work, taking three cross-country flights a year would generate more per-person emissions than a year of heating your home.[70]

Even if you're a "power user" of cars and exceed that mileage, the reason why I've placed "our homes" in front of "our travel" in terms of the impact we can make is because where we choose to live so fundamentally informs how much we travel every day.

But first, why do we even drive in the first place? According to the US Department of Transportation's most recent edition of the National Household Travel Survey in 2017, driving to and/or from work is only the third most common reason for driving, at around 17 percent of all trips. This puts it just ahead of driving for social, recreational, or dining purposes (15 percent). However, it trails shopping and errands (around 20 percent) and "regular home activi-

---

[69] United States Environmental Protection Agency, "Greenhouse Gases Equivalencies Calculator— Calculations and References."

[70] "Google Flights," Google, accessed July 2024, https://www.google.com/travel/flights/

ties" like chores or coming home to sleep (around 34 percent). Most drives are relatively short in length as well, with three-quarters of all drives ending up at under 10 miles.[71]

So outside of relocating where we live, work, or play, what can we do about how we get around? If you are able-bodied, and given that 21 percent of all drives are for only 1 mile or less, walking is always a great emissions-free choice. And because three-quarters of drives are under 10 miles, cycling is the next-best emissions-free option. Biking or walking enough to cut down your annual driving mileage by half, for example, would reduce your emissions from driving enough to equal the same amount of carbon sequestered by over 3 acres of forests in a year. Even better, it can be way more cost-effective over time (no need to buy as much gas!).

Plus, with the emergence of e-bikes, biking longer distances is becoming even more doable (especially with more cargo room for doing errands!). German manufacturer Bosch, which makes e-bike parts, conducted a lifecycle assessment study and found that, on average, e-bikes emit only 5 grams of carbon dioxide per mile—over *fifty* times less than a car.[72]

So if you already have an e-bike, use it! And if you don't have one yet and need to buy one, recall the earlier point about how manufacturing a product is generally far more harmful than its usage. Though e-bikes are more intensive to manufacture than regular bicycles, you'll nevertheless offset the entirety of the emissions needed to manufacture, package, and ship a new e-bike once you've ridden it about 600 miles instead of driving. Buying a used e-bike can drop that threshold significantly.

In short, **if where you live allows it, biking or walking instead of driving is one of the highest-impact activities you can do to lower your transportation and travel emissions**.

Comparing Transportation Options

But what about trips that are too time-sensitive or too far? Or where there isn't infrastructure to bike or walk? Leaving the discussion of remote working for the next section, let's start with some interesting data about motorized transit options.

---

71 "Popular Vehicle Trips Statistics," National Household Travel Survey, US Department of Transportation & Federal Highway Administration, accessed July 2024, https://nhts.ornl.gov/vehicle-trips.

72 "Sustainability at Bosch eBike systems," Bosch eBike Systems, accessed July 2024, https://www.bosch-ebike.com/us/sustainability.

The Congressional Budget Office has calculated something called the "average carbon dioxide emissions per passenger mile," which is effectively a measure to understand how much it takes to move one passenger 1 mile. Based on 2019 data and averages, driving is the least efficient mode of transportation, followed—very surprisingly—by bus, then air, and then trains (passenger trains first, then subway or light-rail).

AVERAGE CARBON DIOXIDE EMISSIONS PER PASSENGER-MILE BY MODE OF TRANSPORTATION, 2019

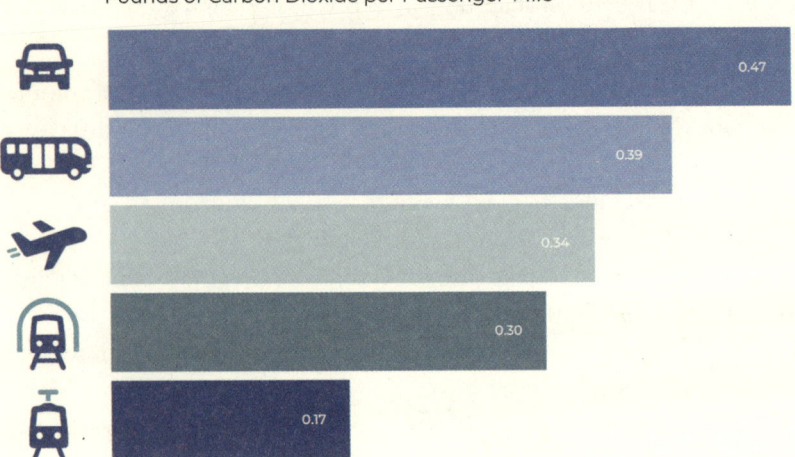

Source: "Emissions of Carbon Dioxide in the Transportation Sector," Congressional Budget Office, https://www.cbo.gov/publication/58861.

But how could a bus be less friendly to the environment than a plane, you ask? Intuitively, we all would expect a bus to be one of the more environmentally friendly ways to get around. And yes, you would be correct—if, and only if, people actually ride those buses. Sadly, in America, because most buses are operating at a fraction of capacity most of the time, this number is far higher than it should be. In the UK, where buses are utilized more actively, the BBC reported in 2019 that buses (specifically coach buses traveling long distances) can be even more carbon-friendly than domestic rail.[73]

However, the wrong conclusion to make is that you should never ride a bus. In fact, it's the opposite. **If transit options exist for you that would run irrespective of whether you're in the vehicle, you should always prioritize those options.** That way, you would be avoiding the need to op-

---

73 "Climate Change: Should You Fly, Drive, or Take the Train?," BBC, August 23, 2019, https://www.bbc.com/news/science-environment-49349566.

erate a new vehicle while helping *lower* the carbon dioxide per passenger mile of that mode of transit! So if a bus, train, or car pool is available to you to get you where you need to go, it's always the right choice for the environment over driving there, as your effective additive carbon emissions would be zero. Bonus points if you can walk or bike there instead.

One caveat here is that "induced demand" is a very real problem when looking at the activities of many people at once. You riding a half-empty bus does not induce more demand, since you're just helping to maximize capacity. But you and everyone else buying more plane tickets than plane seats available can generate the demand for a new plane—and more emissions. This is just to make sure your takeaway from this isn't "always fly because planes are always flying anyway."

> Using alternative means of transportation such as a bicycle, bus, or train is generally more sustainable than driving.

Gas-Powered Vehicles versus Electric Vehicles

Another caveat is that not all cars are created equal. While you are almost always better suited to reduce your impact (both on the planet and on traffic) by using public transit, as your car gets better fuel efficiency and better miles per gallon, you can make driving a bit less harmful—especially if you drive frequently with multiple passengers or if you drive much more efficient vehicles like motorcycles or mopeds.

There are also a lot of things you can do to help your existing car get better fuel mileage by how you drive and maintain it. Per the EPA, you generally use more fuel to accelerate quickly rather than gradually and lose more energy by braking quickly. Cruise control (and driving more slowly in general) are great ways to control this. Further, keeping your vehicle well-maintained—from keeping tires inflated, to keeping the exterior clean, to removing unnecessary weight and items from the interior—can contribute to better fuel economy.[74]

If you're an owner of an EV (and especially if you live on a grid with better access to renewable or low-carbon energy), you can reduce impact even further. Many EVs will have a measure called MPGe, or **miles per gallon equivalent**, to estimate the relative amount of mileage you get compared to

---

[74] "Your Mileage May Vary," United States Environmental Protection Agency, last modified February 8, 2024, https://www.epa.gov/greenvehicles/your-mileage-may-vary.

a gas vehicle. I own a plug-in hybrid vehicle that's currently telling me that my lifetime MPGe is 72, or three times the fuel efficiency of the average car in America. And though you (almost) always benefit from having a car with higher miles per gallon (regardless of whether they're electric), I mention EVs as being most beneficial because their MPGe is almost always significantly better than the miles per gallon of most gasoline-powered cars, including hybrids.

That being said, I get a lot of questions about whether EVs truly are better for the environment than gasoline-powered vehicles. Most often those questions focus on the raw materials that go into the batteries and how to recycle those batteries after the car drives its last mile. While the recycling question is a fair one (thankfully, many new start-ups that focus on battery recycling specifically are emerging), one way to answer the question regarding emissions is to go back to our trusty friend known as the lifecycle assessment to figure out whether manufacturing an EV is worse than manufacturing a comparable gas vehicle.

In a 2021 study, the EPA and the DOE looked at the lifetime emissions of two similar-sized cars. The EV is assumed to have a 300-mile range and charges using average US grid emissions, while the gas car is assumed to have 30.7 miles per gallon (better than average). Each is assumed to have a lifetime usage of 173,151 miles.[75]

As you can see in the following figure, generally the "emissions per mile" of an EV is less than half of a gas vehicle using the assumptions above. Keep in mind that even if you're not using electricity to power a gas vehicle, you are still using electricity to extract and process the gasoline before it even gets to your car. At the same time, it is not incorrect to notice that the emissions it takes to manufacture an EV and its battery are larger (over 50 percent) than those of a comparable gas vehicle.

---

[75] "Electric Vehicle Myths," United States Environmental Protection Agency, https://www.epa.gov/greenvehicles/electric-vehicle-myths.

WHAT WE CAN DO

LIFECYCLE EMISSIONS FOR AN ELECTRIC VEHICLE AND A GASOLINE CAR

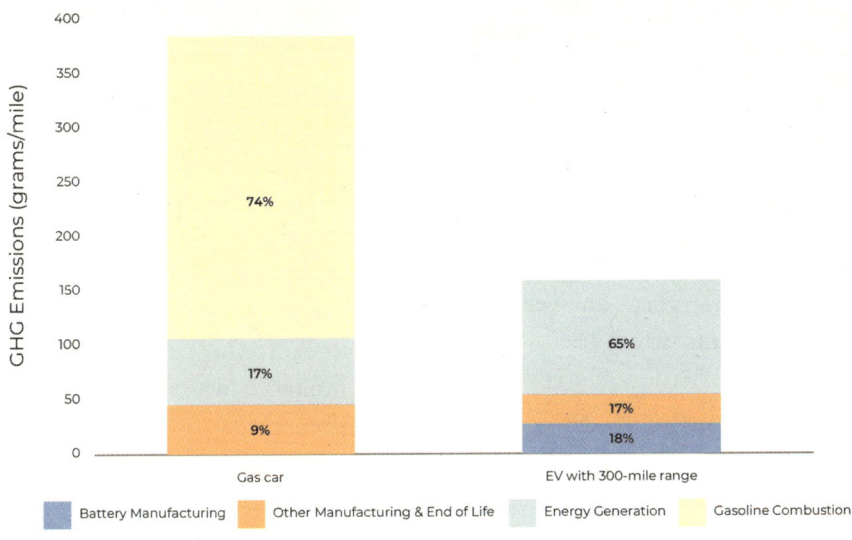

Source: "Electric Vehicle Myths," United States Environmental Protection Agency, https://www.epa.gov/greenvehicles/electric-vehicle-myths.

In fact, the manufacturing emissions of an EV roughly equate to driving a gas vehicle 35,000 miles. This leads to another counterintuitive, yet data-supported, realization: If you have fewer than 35,000 miles left on your gas car before you discard it, you are better off driving your existing gas vehicle than buying a new EV and inducing the demand to manufacture a new car. In a cruel irony, somebody who immediately bought a new EV to "be more sustainable" rather than waiting until they needed (not wanted) to buy a new car might have done the less sustainable thing!

This helps to underscore one of the tenets of the book. Because it just takes so much waste to make things, **you are almost always better off using something you already have rather than buying something new, even if the new thing is more energy- or fuel-efficient**.

Right now, you may be asking, "I don't need a new electric vehicle. What about if I buy a used one instead?" Yes, buying secondhand or used (i.e., the "Reuse" part of "Reduce, Reuse, Recycle") is generally a much better option than buying something new. However, cars (and other critical goods) are a bit of a trickier category when it comes to secondary markets. For clothing, for example, something you buy at Goodwill is likely to be thrown away if not

> **Electric vehicles produce fewer lifetime emissions than gas-powered vehicles but take two or three years of driving to realize that benefit.**

purchased, and the person who originally donated their old shirt likely has other existing shirts they can wear.

The secondary market for cars, on the other hand, doesn't operate quite like this. Contrary to used clothing, demand for used cars often outstrips supply, so buying a used car likely doesn't prevent it from going to waste. Further, the person who sold you the car (unless it was an extra car or they plan to stop driving entirely) likely needs to buy a new car to replace it. So while buying a used car may mean that you didn't directly pay an automaker to make a new one, you might have paid them indirectly and have still "induced" the manufacture of a new car.*

*For the free market enthusiasts out there, I recognize that induced demand for electric vehicles CAN send a market signal to both auto manufacturers and car dealerships that EV development and promotion deserve investment, which can lower costs and increase accessibility to EVs to the middle and working classes. This is harder to measure but a very real thing to consider, though I am not convinced that delaying an EV purchase by two or three years makes a meaningful difference in a massive corporation's willingness to invest in this long-term market shift.

In short, **if you want to lower your impact, minimize (i.e., reduce) the amount of transit you need to do**. If you need to go somewhere, public transit, walking, or biking are always your best bets for getting around town. If you need to drive, carpooling in an EV is your best option—unless you don't have one yet, in which case you should drive your existing gas vehicle 35,000 more miles first (i.e., reuse). Unless, of course, you're a free market economist, and you think it's most important to send a market signal to auto manufacturers *right away* that EVs are the future to spur more immediate corporate investment in them versus legacy gasoline vehicles. Simple, right?*

*I do hope you have given yourself some grace by this point due to how complex sustainability can be. Even as a full-time professional in this space, I am learning new and counterintuitive things as I write this book thanks to the recent emergence of better sustainability data.

WHAT WE CAN DO

Vacations

Before closing out this section on our transportation, I want to leave with a quick segment about vacations. This is typically the "biggest" travel many of us do and one of the bigger sources of emissions for those of us who are fortunate enough to take them.

Most often, the way we vacation in America is to travel. This can mean driving to Florida, taking a cruise to Alaska, flying to New Orleans, or taking an RV to one of our national parks. But as we've covered so far, the farther the distance we travel for our vacations, the worse an environmental impact they tend to have. Just two cross-country flights in a year (e.g., from Los Angeles to New York City) generate around a metric ton of greenhouse gas emissions per person, between a quarter and half of an average person's annual driving emissions. The more we vacation and the farther away we do it, the bigger the negative impact.

Plus, the activities we do *while* on vacation tend to carry a certain amount of emissions with them, such as the energy required to power a chairlift at a ski resort or the emissions from the hotel you're staying at. For hotels specifically, a 2022 study done by the UK's Department for Environment, Food, and Rural Affairs estimated the emissions per night of an average hotel stay in America to be around 16 kilograms, the equivalent of about 40 miles of driving.[76] This ballpark figure was confirmed (via pleasant surprise!) when I recently booked a hotel room through Marriott, which now shares the per-night carbon and water footprint of your stay (which, for me, came to 18.39 kilograms of emissions).

And while you've hopefully turned off the lights at your normal home when on vacation, the amount of persistent energy usage in your house (like refrigerators) means that your hotel stay is adding emissions on top of your home's, not outright replacing them.

Let's cover some mainstream vacation types and how to determine their impact based on the same "one-week vacation" assumption, ranked from "least harmful" to "most harmful."

---

[76] "The Carbon Emissions of Staying in a Hotel," Circular Ecology, February 16, 2023, https://circularecology.com/news/the-carbon-emissions-of-staying-in-a-hotel.

The "Staycation" (0 to 150 Kilograms of Carbon Dioxide per Person)

While not necessarily flashy, the staycation is almost always the lowest-impact way to take time off that you have, since you're ideally traveling locally where you already live and generating only minimal added emissions. Even if you decide to get fancy and book a hotel near home for a week, that plus local transit would only account for up to 150 kilograms of CO_2e (around 2.5 trees) if you're staycationing solo. It drops to 50 kilograms per person if you're with three people.

The Camping Trip (130 to 600 Kilograms of Carbon Dioxide per Person)

One common type of road trip involves using a camper or RV and driving out into nature. A midsized Class C RV typically gets around 15 miles per gallon, compared to around 23 miles per gallon for the average passenger vehicle.[77,78] Furthermore, because it's often powering things like heat, televisions, and stoves inside, an RV uses around 20 kilowatt-hours of electricity per day, which is roughly the equivalent of an extra gallon of gas per day.[79]

On the surface, an RV trip might look to be a less sustainable option because of its lower fuel efficiency and need for electricity. However, the benefit of this kind of trip is that once the RV reaches its destination campsite, it often stays there—and you don't need a hotel room, since you brought yours with you. Let's say that there are two days of driving (each being around 500 miles) and six nights of camping. That would burn around 70 gallons of gas, which equals about 600 kilograms of carbon dioxide, or around ten trees' worth.

However, another benefit of the RV is its capability to fit multiple passen-

---

[77] "What Is the Typical RV Gas Mileage?," Cruise America, accessed July 2024, https://www.cruiseamerica.com/trip-inspiration/rv-gas-mileage.

[78] Sean Williams, "Motorhome Gas Mileage [Definitive Guide]," Topnotch Outdoor, January 15, 2022, https://topnotchoutdoor.com/motorhome-gas-mileage/.

[79] "Things You Need to Know About RV Electricity," Renogy, 16, 2022, https://www.renogy.com/blog/things-you-need-to-know-about-rv-electricity

gers at once. If the same three people from the staycation were to fit in this RV, the per-person emissions would drop to only 200 kilograms of CO_2e per person. Tent camping can bring this number down even further.* Using a regular car for the same type of trip would only generate around 400 kilograms of CO_2e, which is around six to seven trees' worth. Camping with three people brings the per-person rate down further, to 130 kilograms of CO_2e per person for the week.

Another benefit of tent camping is that you're more likely to get there via the vehicle you already own and use frequently. Not included in this calculation is the embodied carbon of an RV, which is considerable for a large vehicle that is used infrequently.

The Road Trip *(200 to 1,200 Kilograms of Carbon Dioxide per Person)*

Next up is the classic American cross-country road trip. Per the EPA, the average fuel economy of cars and light trucks in 2020 was about 23 miles per gallon. Every gallon of gas burned produces around 9 kilograms of emissions, roughly equivalent to charging a smartphone one thousand times.

Let's look at two kinds of road trips, starting with a cross-country road trip. Let's assume that on this road trip, each day of driving is around 400 miles, with six nights of hotel stays added on top (each of which is the equivalent of driving an additional 40 miles). That would generate around 1,200 kilograms of emissions, or about twenty trees' worth. If you bring two additional people along and share hotel rooms, the per-person rate drops to 400 kilograms each.

The other kind of road trip is the "drive to your destination" trip, like driving to a cabin, to the beach, or into the mountains. If we assume the big part of the drive is getting to and from the destination, the all-in miles driven don't exceed 1,000, and a hotel is booked for the week, then the emissions can drop to around 600 kilograms of CO_2e (or ten trees' worth) total, or around 200 kilograms per person if three people travel together.

The Domestic Flight (300 to 550 Kilograms of Carbon Dioxide per Person)

This brings us to flights to a domestic destination. There are many versions of this, ranging from the classic ski trip, to a trip home for the holidays, to visiting New Orleans for Mardi Gras, or to a bachelor or bachelorette weekend in Las Vegas.

Let's assume that for this trip, you'll mostly stay put at your destination. If the departure airport is 30 miles away from your home and the hotel is 30 miles away from the destination airport, the trip starts with around 50 kilograms of CO_2e. Adding in the round-trip flight adds more on top. A round-trip from Cleveland to Miami, as a proxy, generates around 200 kilograms of carbon dioxide per person.[80] And unlike road trips, which can reduce per-person emissions by carpooling, the assumption is that every flight will be more or less full; adding more people doesn't really reduce the per-person emissions materially.

Adding in six days of hotels and various trips around the destination (e.g., to and from the beach) adds 200 kilograms or so, which can be made a little better by sharing a hotel room and carpooling or taking public transit at the destination. In this example, a solo trip would generate around 450 kilograms of CO_2e (about 7.5 trees), which could drop to around 300 kilograms per person for a three-person trip.

However, let's say that you're doing a big ski trip, also starting from Cleveland. To quote a humorously titled article in French that I found on this topic, "Is skiing an ecological horror?"[81] Given that a flight from Cleveland to Denver still drops you 120 miles away from the famous ski resort town Vail, you still need six nights of stays at hotels that require more heating than the average hotel (because it's cold), and you're riding a ski lift up the mountain all day, the impact adds up. The round-trip flight (about 240 kilograms), the driving (around 100 kilograms), the hotel (about 200 kilograms), and the chairlifts (around 10 kilograms) end up closer to 550 kilograms for a solo trip (or 350 kilograms per person, if shared between three people).

An interesting note here: While the "minimum possible emissions" from a domestic flight vacation are higher than some of the road trip options listed earlier, the "maximum possible emissions" from flying can be *lower* than a really long road trip. This reinforces one of the earlier points in this chapter that

---

80 "Google Flights," Google, accessed July 2024, https://www.google.com/travel/flights/.

81 Thomas Wagner, "Is Skiing an Ecological Horror?," Bonpote, last modified March 3, 2023, https://bonpote.com/en/is-skiing-an-ecological-horror/.

driving (especially if alone) can actually be worse for the environment than flying.

The International Vacation (1,200 Kilograms of Carbon Dioxide per Person)

The next "worst" vacation from an environmental perspective is the international vacation, mostly due to the sheer distance one must fly to get there.

Let's assume you're taking a direct flight from Chicago to Paris for a week in the City of Love. The round-trip flight ends up emitting around 1,000 kilograms of CO_2e per person. If you add in the drive to and from the airport, six days of hotels, and various trips around the city, you're likely adding on 200 kilograms or so of emissions. Of course, if you share a room and carpool or take transit together, you can drop the per-person rate by one-third for the time within the city.

Assuming you and two others go to France together, that could end up emitting over 3 metric tons of CO_2e—over fifty trees' worth. Now, obviously there is a lot of variance to this, as some international destinations are farther away than others, and what you *do* at that destination can vary wildly. But generally, international trips tend to be pretty tough on the planet.*

All that being said, this might be a nice moment to interject an optimistic development when it comes to air travel. Thankfully, many airlines (as well as my own employer!) have started investing in something called "Sustainable Aviation Fuel" (SAF) as an alternative to traditional jet fuel. This new fuel—made of anything from used cooking oil to municipal biowaste—can lower the lifecycle emissions of a flight's fuel by up to 80 percent and make air travel much less taxing.[82] I recently took a flight on Lufthansa and was pleasantly surprised to see the airline give me the option to fully match my fuel usage with SAF.[83] While it cost me almost €800 to do so (yikes!), the cost should come down as the market for this new jet fuel matures and it becomes more widely available. Travelers with access to this fuel option should take advantage if available and financially feasible.

The Cruise (1,300 Kilograms of Carbon Dioxide per Person)

Outside some outlandish vacations not covered here (e.g., chartering a private

---

82 "SAF 101", Roundtable on Sustainable Biomaterials, accessed August 2024, SAF-now.org

83 "Make Your Air Travel More Sustainable Now," Lufthansa, accessed August 2024, https://lufthansa.compensaid.com/

plane to fly around the world), the worst type of vacation you can take for the planet is a cruise. Many aspects of this have already been well documented, such as the $60 million plus in fines that Carnival Cruise Lines has paid in the last decade for "deliberate pollution" of plastic trash and oily discharges into the ocean.[84] Despite this, a 2023 survey by *Time* magazine found that 50 percent of respondents thought that cruise vacations were actually eco-friendly![85]

In terms of pure greenhouse gas emissions, cruises also take the cake. A 2019 study for the Port of Seattle found that 96 percent of all cruisegoers departing from Seattle (often for a cruise to Alaska) had to travel to Seattle in the first place, with 85 percent of them flying there (the plurality coming from southern or other western states, like California).[86] So to even get to and from the cruise, you're already burning around 400 kilograms of CO_2e per person, including transit to and from the airport (using an origin point of Dallas as an example).

Then, for the cruise itself, while detailed emissions data is not widely available, the International Council on Clean Transportation estimated in 2017 that cruise ships emit more than twice the emissions per mile than airplanes.[87] A round-trip cruise from Seattle to Juneau is around 2,000 miles of travel; at a rate of 400 grams of CO_2e per person per mile for a "more efficient" large cruise ship, that generates 800 kilograms from moving, plus the additional 100 kilograms (approximately) from emissions associated with your stay on the boat itself.

All in, this kind of cruise can generate around 1,300 kilograms of CO_2e per person—more than an international trip to Paris and the per-person equivalent of over twenty trees, each grown for ten years. And surprisingly, it is likely that smaller vessels are *even worse* in terms of pollution. While the smaller boat itself will generate fewer emissions and waste overall, because you can't fit as many people on it, the per-person emissions are likely higher than on a very large ship, which can fit many people at once. For example, smaller expedition companies such as Oceanwide Expeditions, which does cruises to the Arctic and Antarctica, have acknowledged that their per-passenger emissions

---

84 Merrit Kennedy and Greg Allen, "Carnival Cruise Lines Hit with $20 Million Penalty for Environmental Crimes," NPR, June 4, 2019, https://www.npr.org/2019/06/04/729622653/carnival-cruise-lines-hit-with-20-million-penalty-for-environmental-crimes.

85 Alejandro de la Garza, "The Cruise Industry Is on a Course for Climate Disaster," *Time* (online), June 13, 2023, https://time.com/6285915/cruise-industry-climate-action-emissions-passengers/.

86 McDowell Group, *Port of Seattle: Alaska Cruise Passenger Survey 2019*, October 2019, https://www.portseattle.org/sites/default/files/2019-10/Alaska%20Cruise%20Passenger%20Survey%202019.pdf.

87 Brian Comer, Ph.D., "What If I Told You Cruising Is Worse for the Climate Than Flying?," International Council on Clean Transportation, May 16, 2022, https://theicct.org/marine-cruising-flying-may22/.

WHAT WE CAN DO

> **Vacationing locally, camping, and avoiding cruise ships or international flights tend to help minimize the emissions of your vacations.**

could indeed be lower but come as a trade-off for enabling more climate research opportunities.[88]

To their credit, many cruise lines have started to make commitments to decarbonize the cruising industry, signing up to things like the Glasgow Declaration on Climate Action in Tourism. And hopefully, if you're reading this book a decade from now, much of the above might no longer ring true. But until then, cruise ship vacations should be avoided if possible if your interest is to minimize the impact of your vacations.

---

[88] "The Impact of Small vs. Large Cruise Ships," Oceanwide Expeditions, accessed July 2024, https://oceanwide-expeditions.com/blog/the-impact-of-small-vs-large-cruise-ships.

Type of Vacation	Emissions Estimates Based on Size of Party (kg of CO_2/person)		
	Solo Trip	Two People	Three People
Staycation	150	100	50
Tent Camping Trip	400	260	130
Domestic Flight Trip	450	375	300
Ski Trip	550	475	350
RV Camping Trip	600	400	200
Destination Road Trip	600	400	200
Cross-Country Road Trip	1,200	800	400
International Vacation	1,200	1,133	1,067
Cruise	1,300	1,250	1,200

Discussion Points

1) Think about the types of trips you make most frequently. What is preventing you from taking alternative modes of transportation to get there? What could solve this?

2) EVs produce fewer emissions than gas vehicles but require critical raw materials for their batteries that gas vehicles don't. How should this tension be balanced?

3) What false assumptions do you think have contributed to people's perception that cruising is a more eco-friendly vacation type despite it being one of the worst? How can we combat these false assumptions?

4) What do you value more in a vacation: the journey or the destination? How does that affect your choices from an emissions standpoint?

Chapter 5: Your Purchases, Diet, and Stuff (28 Percent of Your "Personal" Emissions)

The next set of impacts from how you live your life ties in closely to the decisions you make most frequently (e.g., "What kinds of stuff do I buy?" "How do I use it once I've bought it?"). Experientially, this is often when we are most consciously faced with whether our actions "feel" sustainable. Things we buy in stores lean into this and increasingly tout their various eco-friendly characteristics.

Unfortunately, each category of thing you may buy (e.g., food, clothes, appliances, medicine) operates with its own rules and contexts for what constitutes being more or less sustainable. However, we can carry with us the earlier framework of "Make It, Move It, Use It, Lose It"; by applying it—specifically, by remembering that making something is often more wasteful than using it—this can help us create a helpful framework to qualify the decisions we make at the store and at home.

This will be helpful to reduce impact in two ways. In addition to helping reduce your waste in your personal life, every purchase you make as a consumer sends a demand signal to the companies and retailers selling those goods. It might seem inconsequential, but in aggregate these purchase decisions have major ramifications for how the product's makers envision their product roadmaps. Enough people voting with their wallets (which doesn't always mean buying a more expensive "eco-friendly" product) can persuade business executives to greenlight the pursuit of more sustainable goods.

I have seen this in my own career, and many studies, like one from global consulting leader McKinsey & Company, have found that consumers "care about sustainability—and back it up with their wallets."[89] One environmental weakness of the capitalistic society we live in—where companies generally seek higher profits at the expense of everything else—can actually be exploited for the good of the planet by aligning a company's profit incentive with delivering more sustainable products. Even the most begrudging company would be

---

[89] Jordan Bar Am, Vinit Doshi, Anandi Malik, and Steve Noble, "Consumers Care about Sustainability—and Back It Up with Their Wallets," McKinsey & Company, February 6, 2023, https://www.mckinsey.com/industries/consumer-packaged-goods/our-insights/consumers-care-about-sustainability-and-back-it-up-with-their-wallets.

foolish to ignore sustainability if it means they'd be leaving money on the table from their customers. (Though greenwashing is something to keep an eye out for, which we'll discuss later.)

Before I list guidance for some of the major categories of purchases, I want to stress that **there is no way today to make the "perfect" choice every time**. We simply don't have enough data from the brands we buy from to make a truly informed choice every single time—and we certainly don't want to pull up a company's eighty-page sustainability report before every purchase. But that is OK! We should not let "perfect" be the enemy of good. And most importantly, we should make decisions that make sense for us in the context of our own lives. A lot of people making a little bit of a difference in their daily lives has way more impact than very few people making significant changes.

Your Food (19 Percent of Your Emissions, Plus 3 Percent for Cooking and Storing It)

Let's start with the single thing we all buy the most: food! The agriculture industry is responsible for around 10 percent of emissions and 40 percent of the water usage in America each year, and it's one of the more immediate places where you can have a meaningful impact—in terms of reducing emissions as well as, in some cases, animal well-being and other social causes.[90] Per the EPA, given America generated 664 million metric tons of emissions from agriculture in 2022, then on a per-person basis, we each account for roughly 2 tons of carbon dioxide from our diets annually.[91]

What It Takes to Make Our Food

As we saw in the introduction, food takes a *ton* of emissions, waste, and water to create. The example we used earlier was for eating twenty burgers—the emissions equivalent of burning down a tree and drinking over fifty years' worth of water. And that's not even considering that almost all food, whether in a grocery store or a to-go box, is packaged in at least some amount of nonrecyclable plastic.

So given this, the first and easiest thing you can do is to **eat all your food**. In America, the United States Department of Agriculture (USDA) estimates that *between 30 and 40 percent of all food is thrown away*.[92] Think of all the times

---

[90] United States Geological Survey's Water Availability and Use Science Program, *Estimated Use of Water*.
[91] United States Environmental Protection Agency, *Inventory of U.S. Greenhouse Gas Emissions and Sinks*.
[92] "Food Waste FAQs," US Department of Agriculture, accessed July 2024, https://www.usda.gov/food-waste/faqs.

that you've had to throw out moldy or spoiled food over your life because you didn't eat it. At scale, we would need 30 percent less land, use 30 percent less water, and emit 30 percent fewer emissions if we actually ate everything we produced. In our daily lives, this mostly means being smarter when we shop to reduce the overall amount of food we buy by purchasing food that we have a clear plan to cook or eat before it expires. And if you're finding yourself throwing out a lot of food after you cook it, consider making smaller serving sizes!

Other than actually eating your food, generally reducing your red-meat intake (beef in particular) is one of the most impactful things you can do in your daily life. Today, per the USDA, beef and dairy production accounts for a staggering 44 percent of all American farmland usage and around 18 percent of all American land usage, period.[93] People often forget that a lot of the "non-meat" food we grow also goes to feeding livestock. Per the USDA, livestock consumes about 40 percent of all corn grown in the United States. All of that land being used for cattle (and its food) could otherwise be used for more natural ecosystem development.

> Minimizing food waste reduces the emissions associated with our diets.

---

93 US Department of Agriculture, *2017 Census of Agriculture: Highlights—Farms and Farmland*, 2017, https://www.nass.usda.gov/Publications/Highlights/2019/2017Census_Farms_Farmland.pdf.

HOW AMERICA USES ITS LAND

Source: "Here's How America Uses Its Land," Bloomberg, https://www.bloomberg.com/graphics/2018-us-land-use/.

People often ask me about how they can help fight deforestation. If you look at the causes of deforestation in the Amazon rainforest, it's not to cut down trees to produce timber. It's primarily to make space for pastures to raise cattle (or to grow food for the cattle). Buying more beef—even if that beef is more local—can induce more demand for it globally. (So if you're looking for a way to help save the Amazon, eating less beef is a great place to start!)

Fish can also be incredibly problematic, especially if that fish is being harvested in the wild. You may have heard of the Great Pacific Garbage Patch, a floating island of trash in the Pacific Ocean that is, at the time I am writing this book, around *twice as large as the state of Texas* (and three times the size of France!). Research published in the journal *Scientific Reports* by the Ocean Cleanup in 2022 estimated that over 75 percent of this waste came from discarded commercial fishing plastic, like fishing nets and ropes.[94] If you're buying seafood, chances are whoever fished that seafood just threw away their nets into the water once they caught your fish.

And as most fishing (unlike other types of foods) involves the capture of wild animals, overfishing is immensely dangerous to global biodiversity. An

---

94 The Ocean Cleanup, "Over 75% of Plastic in Great Pacific Garbage Patch Originates from Fishing," September 1, 2022, https://theoceancleanup.com/press/press-releases/over-75-of-plastic-in-great-pacific-garbage-patch-originates-from-fishing/.

WHAT WE CAN DO

oft-quoted study that came out in 2014 by the Ellen MacArthur Foundation predicted that by 2050, there will be more plastic in the ocean than fish—a combination of both increased plastic production and overfishing.[95] And per the Food and Agriculture Organization of the United Nations in 2019, 35 percent of all marine fish stocks are overfished—compared to just 9 percent in 1978—and only 7 percent are underfished.[96] Luckily, the rate of aquaculture (i.e., fish farming) has grown considerably in comparison to traditional marine fishing. So if you're eating fish, a more sustainable option is to buy farmed or locally caught fish.

Thankfully, for those of us who quite enjoy the taste of meat and don't want to cut it out entirely, there are a number of far less impactful—and almost as tasty—meat alternatives now on the market, from Beyond Burger to Impossible Foods, Field Roast to Gardein, and many more. The same goes for cow milk alternatives, which are typically lower-emitting in comparison.

Interestingly, while converting to vegetarianism is certainly one way to cut impact, it is not the most impactful. Cheese, for example, has a higher carbon and water footprint than chicken. And many vegetarian staples (e.g., most nuts, especially almonds and cashews), though they have a much lower carbon footprint than meat, are incredibly thirsty, requiring more water than every type of meat except for beef. Per WaterCalculator.org, almonds and chocolate require *more* water than beef does to produce—around 500 gallons of water to produce just 4 ounces.[97]

> Reducing red meat consumption reduces your impact on the planet.

And then there is meat that is actually *beneficial* to eat, such as venison that you or a friend may have hunted yourself to keep the deer population in check from overgrazing vegetation. But generally, a vegan diet that gets its proteins from tofu, lentils, chickpeas, and beans will be the most reliable way to minimize the carbon and water impact of our diets.

---

95 World Economic Forum, *The New Plastics Economy: Rethinking the Future of Plastics,* January 2016, https://www3.weforum.org/docs/WEF_The_New_Plastics_Economy.pdf.

96 Food and Agriculture Organization of the United Nations, *The State of World Fisheries and Aquaculture 2022: Towards Blue Transformation,* 2022, https://doi.org/10.4060/cc0461en.

97 "Water Footprint of Food Guide," Water Footprint Calculator, accessed July 2024, https://watercalculator.org/water-footprint-of-food-guide/.

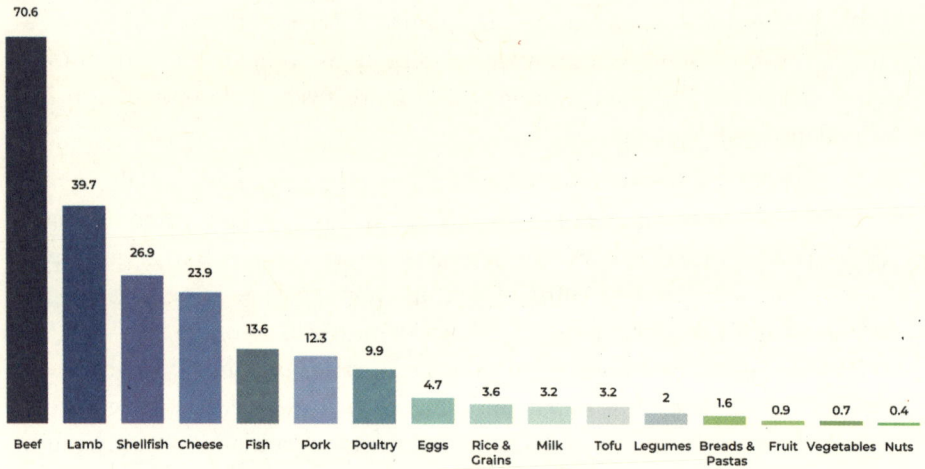

Source: "Food and Climate Change: Healthy Diets for a Healthier Planet," United Nations, https://www.un.org/en/climatechange/science/climate-issues/food.

What It Takes to Move Our Food

This is where the "buy local" movement becomes really meaningful. In addition to helping support your local community of farmers and restaurants, it means that you are likely minimizing the amount of emissions needed to get your food to you in the first place.

Let's say you are at a restaurant in Wisconsin and are given a menu with two types of fish to choose from: walleye or salmon. Which fish was more likely to have been caught locally? For those who don't know Wisconsin very well, it is in the Midwest, far away from any oceans where salmon tend to live. Walleye, on the other hand, is a freshwater fish found in many local lakes. It's a pretty fair bet to assume that it took far less fuel and emissions to get that walleye to your plate than a salmon from a thousand or more miles away.

This is one of the reasons I love farmers' markets so much. They are such great resources not just to buy hyper-locally but also to get educated on what types of foods come from the land around you (and when they are in season). The seasonality portion of the equation also carries impact. If you're buying something that is out of season, it is either being imported from somewhere else, or it's being stored or frozen, which itself takes emissions to run the facilities that the food is stored in.

WHAT WE CAN DO

Seasonality is also relevant in terms of the energy it takes to cook your food. Using your oven for baking during the winter is a great time to do it, as the ambient heat coming from your oven can also warm up your home, as we talked about earlier. But using your oven during the heat of summer means that you are making your AC work even harder to keep your home cool. So changing what you eat—and how you cook it—to follow the seasons can be a nice way to reduce the impact of how you eat your food.

In short, to minimize the impact of your meals, **avoid red meat and wild marine fish, buy local, and cook seasonally**.

> Planning meals with local, seasonal ingredients reduces the impact of your diet.

Discussion Points

1) What do you think makes it so difficult to reduce our red-meat intake? What could be some solutions to help transition your (and our) consumption to lower-impact proteins?

2) Lack of cooking confidence can inhibit a person's capability to plan their diets. How would a focus on better, healthier cooking skills tie in with more sustainable outcomes?

Your Electronics (4 Percent of Your Emissions)

After food, the next biggest category of impact from your somewhat-regular purchases likely comes from your electronics: your TV, sound systems, computers, video games, electric musical instruments, and the like. E-waste, which refers to electronics that end up in landfills, is the fastest-growing component of the municipal waste stream in the United States, per the EPA.[98] Unlike many

---

98 "Helping Communities Manage Electronic Waste," United States Environmental Protection Agency, June 1, 2021, https://www.epa.gov/sciencematters/helping-communities-manage-electronic-waste.

other types of waste, e-waste in particular contains a lot of toxic chemicals that can seep out into the environment, making it potentially a more harmful type of waste than others. (This is part of the reason why things like batteries have an explicit "do not put into the trash" label on them.)

Battery-Powered Electronics versus Plugged-In Electronics

For electronics, an interesting dichotomy emerges as to whether it takes more energy to make it than to use it via our "Make It, Move It, Use It, Lose It" framework. For an individual piece of electronics, the answer generally comes down to whether it runs on battery power or is perpetually plugged into an outlet.

If it is built to run on battery power (e.g., laptops, tablets, phones), it is a far more energy-efficient device, with most of the emissions coming from manufacturing. The reason? Battery life is a competitive differentiator for mobile electronics! Companies have had to treat energy as a scarce resource for mobile electronics and therefore have done amazing work to reduce energy draw. The amount of energy it takes to drive 1 mile in a gas-powered vehicle is the same amount of energy to charge your smartphone for almost *two years*. Recall our earlier example of a MacBook Pro: It would take you over *a decade* of regular usage before the emissions from using and charging the device exceeded the energy to manufacture it.[99]

Plugged-in electronics, on the other hand, assume that energy is always available. They have had less of a driver to optimize for energy efficiency (though there are still *some* reasons, like achieving an Energy Star rating or preventing the device from overheating). Plugged-in electronics nowadays, especially video game consoles, are also often "on" 24/7 (usually in some version of a sleep mode) to provide users with a faster boot-up experience. It'd be like keeping a convertible vehicle in your garage running all the time so that when you get in the car, you can drive away ten seconds faster than if you had needed to turn the ignition.

> **Battery-powered devices tend to be more energy efficient than plugged-in devices.**

As a result, many plugged-in electronics tend to generate more emissions over their lifetime from being

---

99 Apple, *Product Environmental Report: 14-Inch MacBook Pro.*

used rather than from being made. While specific numbers for televisions (which are typically the largest and most used types of electronics in households) aren't particularly available, Samsung estimated in 2020 that about 30 percent of lifetime emissions of its 55-inch 4K UHD TV came from manufacturing while the rest (70 percent) came from energy to power it.[100]

Reducing the Impact of Your Electronics

From an energy-usage perspective, the average American household uses around 632 kilowatt-hours of energy on electronics each year, roughly the same amount of energy needed to propel a car 700 miles.[101] However, that is not the only source of emissions from your electronics. If you're, say, streaming a show on Netflix or playing a video game online, you are also drawing energy and emissions from data centers and servers—not to mention the emissions that went into making the show or the game!

Unfortunately, emissions data that provide a holistic view between the energy draw of your television and the energy draw of the cloud servers providing your streaming shows have not yet been conclusively determined. This is an evolving and incredibly complex science that needs to calculate both the manufacturing and usage emissions of the electronics in your home with the manufacturing and usage emissions of servers and data centers that are often streaming the content to your devices.

To its credit, Netflix (in partnership with the Carbon Trust in 2021) published a paper attempting to get a first view of the carbon impact of streaming on-demand content, albeit only for the energy of the devices and servers and not yet the manufacturing impact. They found that, in Europe, every hour of watching Netflix generated about 55 grams of CO_2e.[102] In other words, watching your favorite show over and over for seven hours is, impact-wise, roughly the same as driving your car about 1 mile.

However, the impact of watching or gaming scales pretty consistently with the *size* of the screen you're watching it on. Watching Netflix on a 50-inch TV may be closer to 55 grams of carbon dioxide, but watching it on a laptop is only 16 grams (around 3.5 times smaller), and viewing it on a smartphone is only 8 grams (around seven times smaller!). You could watch *one hundred different three-hour movies* on your smartphone, and the impact would still be lower

---

100 Samsung, *Life Cycle Assessment for Display Products*, 2020, https://image-us.samsung.com/SamsungUS/epeat/2020_Life-Cycle_Assessment_for_Display_HHP_F_1109.pdf.

101 US Energy Information Administration, "Table CE5.3b, Detailed Household Site Electricity End-Use Consumption, Part 2—Averages, 2020," last modified March 2024, https://www.eia.gov/consumption/residential/data/2020/c&e/pdf/ce1.1.pdf.

102 Carbon Trust, *Carbon Impact of Video Streaming*, June 2021, https://ctprodstorageaccountp.blob.core.windows.net/prod-drupal-files/documents/resource/public/Carbon-impact-of-video-streaming.pdf.

> Watching movies and television shows on a smaller screen is lower impact than watching them on a bigger screen.

than the emissions of driving to watch something at a movie theater more than 3 miles away.

For electronics, the word *reduce* ends up meaning three things: **REDUCE the number of electronics you buy (especially if your old ones still work), REDUCE the time your electronics are on, and REDUCE the size of the screen you use for watching!**

If you do need to buy something, first ask yourself, "Does this need to be an electronic in the first place?" There are many new "smart" devices that don't need to be smart at all and just add unnecessary emissions and e-waste to a perfectly good item, like a trash can with a motion-sensing lid versus a foot pedal.

Assuming the electronic is valuable to own, always consider buying it secondhand or refurbished—rather than buying something brand new—to help reduce the waste from manufacturing. If you have the choice, look for smaller electronics or ones that are marked as being more energy efficient (e.g., having an Energy Star certification or showing a smaller number of estimated kilowatt-hours).

And if you're looking to get rid of an old electronic, there are *many* programs out there that will happily pay you to give them back. Looking for tradeback programs by the manufacturer or checking at local electronics stores like Best Buy can often help yield more information. If there isn't a way to trade them in, there are often many specialty electronics recyclers or nonprofits that would happily accept your old electronics. It's just a bit more work since you can't recycle them curbside.

Can Electronics Save You from Creating Emissions and Waste?

Before we shift topics, I also want to make sure we zoom out and provide a broader context as to

> Buying refurbished electronics is more sustainable than buying new electronics.

how electronics can actually *save* you emissions and waste. Throughout this section, we have been making a comparison to the energy draw of electronics versus burning a gallon of gas or driving a certain distance. In almost every case, using electronics at home for entertainment is generally a far more sustainable alternative than driving for entertainment. Playing Xbox with your friends online will almost always be lower-impact than driving to a friend's home. Watching a new movie on your home theater system is likely lower in emissions than driving to a movie theater and back. Some things you can do on your electronics could also help enable sustainable outcomes themselves, like using the Ecosia search engine, which plants trees as you browse the Web, or using a generative artificial intelligence service like Copilot (powered by ChatGPT) to help locate hard-to-find source material to more quickly write books like this one.

So as you think about what electronics to own, if you think it could meaningfully enable you to avoid driving for entertainment, then it may be worth it to avoid far worse energy usage from your car!

Discussion Points

1) Most people use electronics every day. What are some features you wish your electronics had to help you use them more sustainably? What information would you want to see to help you select electronics that were made more sustainably?

2) This section gave an example of watching Netflix being more sustainable than driving to the movie theater. However, society would be far less interesting if nobody ever left their homes. How should we determine if and when it is "okay" to generate emissions?

Your Clothes (2 Percent of Your Emissions)

After electronics, the next set of things you're using (if not buying) every day is clothing—at least, assuming that you are not a nudist. Unfortunately, there aren't a lot of resources available on the lifecycle impact of individual pieces of clothing. The World Bank estimates that up to 10 percent of global emissions come from apparel.[103] But what about the clothes you're wearing right now?

Shoemakers have probably given us the best sense of the emissions impact of our shoes. I own a pair of Allbirds that have the lifetime carbon emissions printed on each shoe. (It's 9.91 kilograms of CO_2e.) For other types of clothing, there are two lifecycle assessments that are most often quoted to describe the total impact of an individual piece of clothing. The first, published by Levi's, focuses on its 501® jeans from 2015 and is almost certainly outdated.[104] The second, funded by Woolmark in Australia and published in the *International Journal of Life Cycle Assessment* in 2020, claims to be the world's first peer-reviewed textile fiber cradle-to-grave lifecycle assessment study, though it's specifically about wool clothing.[105]

Though these are comparing apples to oranges, they generally reinforce a core thesis from this book: **Making new clothes is way more wasteful than wearing and washing clothes**.

In the Levi's example for jeans, Levi's found that, on average, one pair of jeans roughly takes 4,000 liters of water to make and clean over its lifetime, with the majority of this (about three-quarters) coming from cotton and manufacturing.[106] This number can wiggle slightly depending on how frequently you wash your jeans, but even washing them weekly with an old, inefficient washer uses less than twice the water that it took to grow the cotton that went into it. The Woolmark study found something similar: Around 75 percent of water used across the lifetime of wool clothing came from raising the sheep and manufacturing the clothing, with the remainder from washing.[107]

When looking at greenhouse gas emissions, you also get similar results: Around two-thirds of the emissions associated with jeans (per Levi's) came from making and shipping them, with the remaining third coming from the electricity to wash and dry them over their lifetime.[108] Wool, which is generally washed less often and relies on sheep (which are more carbon-intensive than cotton) to generate its materials, sees almost 90 percent of its emissions

---

103 "How Much Do Our Wardrobes Cost to the Environment?," World Bank, September 23, 2019, https://www.worldbank.org/en/news/feature/2019/09/23/costo-moda-medio-ambiente.

104 Levi Strauss & Co., *The Life Cycle of a Jean*.

105 S.G. Wiedemann, L. Biggs, B. Nebel, K. Bauch, K. Laitala, I.G. Klepp, P.G. Swan, and K. Watson, "Environmental Impacts Associated with the Production, Use, and End-of-Life of a Woollen Garment," *International Journal of Life Cycle Assessment* 25 (2020), https://doi.org/10.1007/s11367-020-01766-0.

106 Levi Strauss & Co., *The Life Cycle of a Jean*.

107 Wiedemann, Biggs, Nebel, Bauch, Laitala, Klepp, Swan, and Watson, "Environmental Impacts Associated."

108 Levi Strauss & Co., *The Life Cycle of a Jean*.

WHAT WE CAN DO

coming from making and shipping wool clothing and around 10 percent from wearing it.[109]

So what can we do to reduce the impact of our clothing? It comes down to four key things.

Reduce

In this case, simply buying less clothing—or clothing that lasts longer—is the single most impactful thing you can do. For clothing, the most important thing to avoid is fast fashion. Buying something to wear once and then discarding it is one of the more wasteful trends around today. If you do need to buy new clothing, buying higher-quality, longer-lasting items—or taking a trip to your local Goodwill or other thrift store—is your best bet. Buying used clothing helps reduce impact as well, and both examples save you money over time.

> Buying used clothing is more sustainable than buying new clothing.

Re-Wear

This is an addition to the normal "Reduce, Reuse, Recycle." The more you wear your clothing without washing it (the most impact-laden thing you can do once you own the clothing) can drive down impact significantly. Washing your clothing once a month versus once a week is four times less harmful to the planet. Once you do choose to wash your clothing, your best bet is to wash it in cold water. Levi's estimates that for a conventional washer, this can reduce your annual emissions from clothing usage by around 30 percent. And if you have the option, drying

> Washing in cold water rather than hot water is a more sustainable alternative.

---

109 Wiedemann, Biggs, Nebel, Bauch, Laitala, Klepp, Swan, and Watson, "Environmental Impacts Associated."

your clothing on a rack or a line rather than a dryer can reduce the emissions from wearing clothing by around 50 percent! Put all that together, and washing in cold and line-drying can have a nearly 80 percent lower impact than washing in hot water and drying in a dryer.[110]

Reuse

Once your clothing has officially passed its usefulness to you, your best bet is to donate it or give it away for a second life. Thankfully, there are robust and ubiquitous options for clothing donations across the country. Doing so gives a better chance for someone else to wear your old clothing, helping them avoid needing to buy new and induce the waste associated with making new clothing.

Recycle

Finally, in the case that your clothing is truly no longer wearable, consider ways in which you can recycle it. Could you use old socks or shirts as rags? Could you connect with specialty recyclers who can use the raw materials? Even though clothing isn't curbside-recyclable, there are still options to keep your old clothes in use.

Discussion Points

1) "Fast fashion" has risen in popularity over the last decade—and with it, concerns of its impact. What are some ways that could reduce the impact of fast fashion?

2) What might inhibit people's willingness to buy used clothing? Are there things that can be done to help encourage this?

Your Furniture (1 Percent of Your Emissions)

Another surprising source of emissions and waste for many households comes from furniture, especially as they are often some of the largest (and therefore most waste-laden) items we own. Having lived near a major university for several years, I was always pained to see how many couches and beds got thrown onto the side of the street at the end of a school year once the students were

---

110 Levi Strauss & Co., *The Life Cycle of a Jean.*

ready to move. Especially in the era of the "fast furniture" movement borne from IKEA, anybody who has moved in the last decade has probably faced the dilemma of throwing out perfectly good furniture because it didn't fit the new home or (more commonly) it was "too much of a hassle" to deal with in the move.

Unfortunately, there are not a lot of definitive studies on the carbon footprint of furniture. However, as our "Make It, Move It, Use It, Lose It" framework bears out once again, making furniture—and, in this case, moving it—has an outsized impact because of how big it tends to be and how many materials are needed to produce it. The only reason it's not a higher percentage of our emissions is because of the relative infrequency by which we purchase it. IKEA, in its 2023 sustainability report, estimated about 60 percent of its entire corporate emissions came from sourcing materials and making its furniture, with an additional 13 percent coming from moving the furniture to its stores and our homes.[111] (This lends credence to the idea that moving your furniture is generally more sustainable than throwing it out and buying new furniture.)

Perhaps the most comprehensive study out there today is from a 2011 exploration by the Furniture Industry Research Association. As it is over a decade old, it is in dire need of a refresh, and the numbers should be treated as directionally correct rather than gospel, especially since they are averages over a wide range of different products. But even so, its findings reinforce some of our intuitions: the bigger the furniture, the bigger the carbon and waste footprint.[112] But not always—and this is where material selection becomes so important. Pound for pound, the impact of wood is much lower than alternative materials like foam (the stuff in many couches) and metal.

---

111 IKEA, "The Climate Footprint."

112 Furniture Industry Research Association, *A Study into the Feasibility of Benchmarking Carbon Footprints of Furniture Products*, 2011, https://www.healthyworkstations.com/resources/Environment/FIRA.Carbon-Footprint.pdf.

Item	Estimated Carbon Footprint (kg CO$_2$e)	Roughly Equivalent to... (gallons of gas)
Sofa	90	10
Sofa Bed	88	10
Double Mattress Bed	79	9
Work or Office Chair	72	8
Armchair	43	5
Desk (Rectangular)	35	4
Dining Chairs	27	3
Dining Table	25	3
Bed Headboard	22	3
Bookcase	18	2
Footstool	17	2

So what does this mean for us? First and foremost, we should minimize the amount of furniture we buy and try to keep our existing furniture for as long as possible before swapping it out. And when we *do* need to buy furniture, it's valuable to buy (or even take off the street!) used or secondhand furniture (e.g., through "Buy Nothing" social media groups) to avoid needing to manufacture new furniture.

If you do need to buy brand-new furniture, look for furniture that is made mostly of wood rather than leather or synthetic material (ideally from a manufacturer that has certifications that it's using sustainably forested timber). And, if possible, aim to purchase smaller rather

> Furniture made of renewable resources like wood tend to be more sustainable.

than larger furniture. While everyone loves a big bed, is it *really* necessary for you to own an Alaskan King bed? Or would a full-size bed work just as well?

Discussion Points

1) What are the benefits and challenges of "fast furniture," like from IKEA?

2) A lot of furniture today does not get reused. How could we solve that problem?

Your Other Purchases (1 Percent of Your Emissions)

The final category of stuff we buy is somewhat of a catchall: they're your "other" purchases. This can range from deodorant to plates and laundry detergent. I lump all of these things together, as they are usually the smallest in terms of emissions impact. However, there are certainly opportunities to reduce your impact here, especially for things like waste reduction, where you're likely going through a LOT of plastic packaging within a year. Not to mention that our everyday purchases are probably made of plastic themselves. (Unless it *explicitly* says it's not made out of plastic, just assume that it is. Even many "paper" cups and plates have plastic lining within them.)

Sadly, that plastic is probably not commonly recyclable. Per the EPA, only around 5 to 10 percent of plastic is actually recycled, with the rest sent to landfills or the ocean, where it could take anywhere from twenty to five hundred years to decompose—and often just become microplastics rather than fully disappear.[113] While the individual items may be small, it's worth keeping in mind that the negative environmental impact of even those little impulse purchases will almost certainly outlive you and the generation of your children.

With that uplifting thought out of the way, what should we consider when buying our everyday stuff? Here are a few starters, keeping in mind the general principles of "Reduce, Reuse, Recycle" and that making things is typically more wasteful than using them.

---

113 United States Environmental Protection Agency, *Advancing Sustainable Materials Management: 2018 Fact Sheet—Assessing Trends in Materials Generation and Management in the United States*, December 2020, https://www.epa.gov/sites/default/files/2021-01/documents/2018_ff_fact_sheet_dec_2020_fnl_508.pdf.

Buy in Bulk

If you're confident that you'll be able to use the entirety of the item, it is typically more sustainable to buy in bulk (or to buy the largest version of an item). Two examples would be buying the big bulk packs of toilet paper or buying the "giant size" box of cereal. When things are packaged in bulk, it generally reduces the amount of packaging needed versus buying things individually. Further, buying in bulk may also allow you to make fewer trips to the store, reducing your transport emissions. And as an added benefit, it's usually cheaper!

Bundle Your Purchases

When possible, try to bundle all of your purchases together at once. One "mega day" of errands can reduce your transport emissions to and from the store by avoiding all the return trips that you'd have to make if you made smaller, more frequent trips (unless those smaller trips are by foot or bike). And if you are ordering online, some retailers give you the option of packaging all of your items together at once, minimizing the amount of packaging and trips that they need to make to get things to your door.

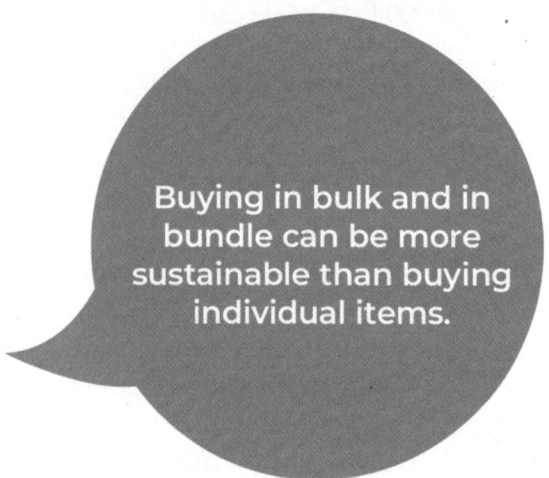

Buying in bulk and in bundle can be more sustainable than buying individual items.

Don't Buy What You Don't Need

I know that I struggle with this one from time to time, but it does bear reminding that the most sustainable purchase is the lack of a purchase altogether. If you don't have an immediate and continued use for something, weigh whether it's worth skipping. And if you're buying things to support an unhealthy habit such as smoking, maybe the benefit to the planet from stopping could help give that extra push you may need. After all, cigarette butts are the most common form of ocean litter, per the National Oceanic and Atmospheric Administration.[114]

---

[114] "What Is the Most Common Form of Ocean Litter?," National Oceanic and Atmospheric Administration's National Ocean Service, accessed July 2024, https://oceanservice.noaa.gov/facts/most-common-ocean-litter.html.

Buy Local Brands

Where possible, try to buy local or state brands rather than national or international ones, with a bias toward products that look to be made in the US. Not only is this great for your local economy, but it also increases the likelihood that you are buying something with fewer emissions associated with the logistics of getting the item to your store (and perhaps less packaging along the journey as well). One product type that fits this guidance particularly well is beer—the more locally brewed, the better!

Buy "Sustainable" Alternatives

Unfortunately, these items can often carry a "green premium" cost-wise. But if you can afford it, try to buy products that market themselves as more sustainable. However, be careful of **greenwashing**, the practice that companies sometimes use to make things sound better for the environment than they actually are. Here are a few things to look for:

- Does their sustainability claim have an asterisk? Make sure to read it. This often means the claim isn't as impressive as the company would want you to think.

- Is their claim vague? If it says "eco-friendly" or "green" with no obvious way to back the claim up, this can draw suspicion.

- Does it have third-party certifications or ecolabels? If it does, that's a great first step, as it means they met a certain environmental threshold to achieve that label.

- Use the "eye" test, especially for packaging. Does it look like it's made without plastic?

Rent Instead of Own

Across all of this, consider whether you even need to own the thing you are buying in the first place. Generally, renting it (or buying secondhand from friends, local zero-waste groups, or secondhand stores) means that fewer new things need to be made in the first place. If you're the second user of an item, it reduces the **per-capita emissions** of making that item by half; if you're the third, then it's reduced by two-thirds, and so on. One great category that works well here is books. Going to your local library rather than buying new books is one of the most sustainable ways to enjoy reading (this book included!).

Bring Your Own Bags

After you've put everything into the checkout aisle, try to bring reusable bags whenever you can. The number of single-use bags you'll avoid using by doing so can get pretty meaningful. And for the times when you forget to bring your bags, consider whether you need a bag at all (the items are likely already in their own packaging anyway) or whether you can get paper bags in place of plastic ones.

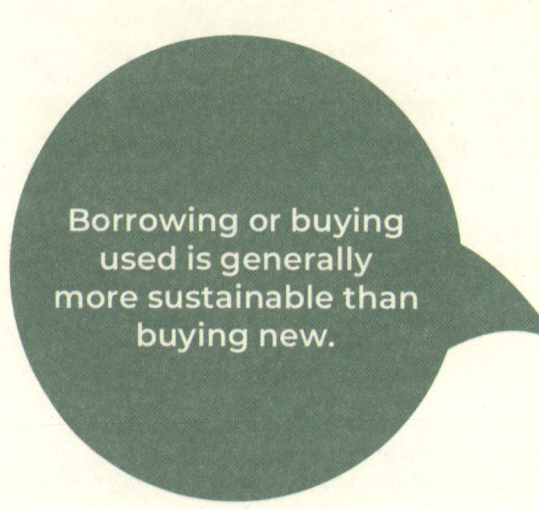

Borrowing or buying used is generally more sustainable than buying new.

Other Considerations

Notice that I did not add "buying at a store" or "buying online" as guidance. The reason for this? It's really hard to say which is more sustainable. In some cases, buying online can be far worse for the environment. If you are doing rush shipping, for example, this often forces your items to come to you via a high-emissions delivery option (often flown to your local distribution center and then driven to your door). **If you're ordering online, the best option is to choose the slowest or cheapest option**, as this implicitly means the delivery can avoid being delivered via air freight. Sometimes it can be an explicit guarantee at checkout, like what my team instituted on Microsoft's website for purchases of PCs or gaming hardware to enable lower carbon delivery options.[115] However, as some online retailers start to electrify their delivery fleets, unless you also own an electric car, it could actually be *more* sustainable to order online. All in all, it depends.

After you've bought your items and brought them into your home, **consider reusing the packaging or bags that you've brought home rather than throwing them away outright**. And if you have access to specialized recycling services, consider whether you can sign up for any that take nontraditional items. I use Ridwell, which allows me to recycle both plastic film and multilayer plastic. Ridwell then recycles these items into things like building

---

115 "Shipping Options, Costs, and Delivery Times," Microsoft store and billing help, Microsoft, accessed July 2024, https://support.microsoft.com/en-us/account-billing/shipping-options-costs-and-delivery-times-63d48e68-3fb8-6d6f-e73f-f321467d6aee.

blocks and drainage material.[116] But generally (and sadly), when it comes to recycling, "when in doubt, throw it out."

Once you're using the items you've brought home, **see what you can do to make them last longer so you can avoid having to make another trip to the store**. I'll give one example that has stuck in my mind for nearly a decade: how to dry your hands with only one paper towel. A man named Joe Smith gave an excellent TEDx Talk on this very topic back in 2012. All you need to do is keep folding the paper towel and reusing it until your hands are dry enough to shake the remaining water off.[117] This talk has gotten around four million views on YouTube by the time of this writing, and I can't recommend it enough.

Discussion Points

1) What "feels" like the least sustainable of your daily purchases? What could be done to make these purchases more sustainable?

2) What could help customers understand the impact of their purchases? What would it take to make that happen?

3) Several brands have a "sustainability facts label" on their packaging that's similar to nutrition labels. Would reading emissions like calories impact the way you buy and use things? What might prevent these from being adopted?

Your Stuff at Home (4 Percent of Your Emissions as Embodied Carbon)

Before closing this chapter, it's worth spending a moment on the things that we have already bought and are around our homes: refrigerators, showers, toilets, ovens, and so on. These and other appliances are things we are theoretically using regularly and rarely replacing. We've already covered some of this in the earlier sections, but it's worth homing in on this a bit more deeply.

---

116 "Multi-Layer Plastic," Ridwell, accessed July 2024, https://www.ridwell.com/pickup-categories/pOXNz0b6.

117 Joe Smith, "How to Use One Paper Towel," TEDx Talk, TEDxConcordiaUPortland, published April 18, 2012, on YouTube, 4:27, https://www.youtube.com/watch?v=2FMBSblpcrc.

Because these things are typically very long-lasting, most of the emissions end up coming from the energy usage of these appliances rather than their manufacture. For example, while industry data is unfortunately scarce, appliance and electronics manufacturer LG estimated that roughly 30 percent of a washing machine's emissions were from embodied carbon, or the emissions required to manufacture it. More power-hungry appliances like refrigerators were estimated to have only 15 percent of their emissions come from manufacturing, with the remainder coming from daily usage.[118] It's important to note that some types of refrigerants used in refrigerators—especially older ones—are some of the most devastating global warming gases around. If a refrigerator is not properly disposed of, these leaks can make its total impact much more harmful.

Per the EIA, if our home happens to have any of the following, we tend to use a significant amount of electricity to power those items.[119] Across all of these items, the EPA estimates the average home's electricity usage annually to be 11,880 kilowatt-hours.

---

118 "Products Application," LG Electronics, accessed July 2024, https://www.lg.com/global/greener-products-application.

119 US Energy Information Administration, "Table CE5.1a Detailed Household Site Electricity End-Use Consumption, Part 1—Totals, 2020," last modified March 2024, https://www.eia.gov/consumption/residential/data/2020/c&e/pdf/ce5.1a.pdf.

WHAT WE CAN DO

Item	Estimated Energy Usage (kWh/year)	Roughly Equivalent to... (miles driven)
Water Heating (at Home)	2,706	3,000
Electric Space Heating	2,484	2,750
Pool Pumps	2,384	2,650
Air Conditioning	2,318	2,550
Hot Tub Heaters	1,479	1,650
Dehumidifiers	999	1,100
Refrigerators	839	950
Clothes Dryers	680	750
Lighting	654	700
TVs and Related Equipment	632	700
Ceiling Fans	226	250
Humidifiers	139	150
Dishwashers	112	100
Microwaves	107	100
Washing Machines	69	50

From an energy point of view, since these items are typically already established in many homes and are not easily or frequently replaced, **the general guidance is to follow two signposts to minimize impact: reducing the *intensity* and *frequency* of the usage**.

Intensity and Frequency of Usage

For things like water heating, AC, or lighting, use these items in a less intense way whenever you can. When taking a shower or washing your hands, for

> Using cold water is significantly more sustainable than using hot water.

example, can you use slightly less hot water? When using AC, can you keep it to 75 degrees Fahrenheit rather than 70 degrees? And when you are using lights, do you need to have *all* of them on? Also, since light bulbs are replaced with decent-ish frequency, what about switching to energy-efficient LED light bulbs, which are less intense than incandescent ones?

How often do you actually need to use your appliances? Can you only power your hot tub (if you have one) for the coldest winter months rather than keeping it heated year-round? Can you shower every other day? Or run the dishwasher less frequently by only doing full loads?

Water Usage

These two concepts also apply to our water usage. In 2016, the Water Research Foundation estimated that, in an average home, 24 percent of our indoor water usage came from our toilets, 20 percent from our faucets, 20 percent from our showers, 16 percent from our clothes washers, and 13 percent from leaks. Surprisingly, only 2 percent of water usage came from dishwashers.[120]

---

120 Water Research Foundation, *Residential End Uses of Water, Version 2*.

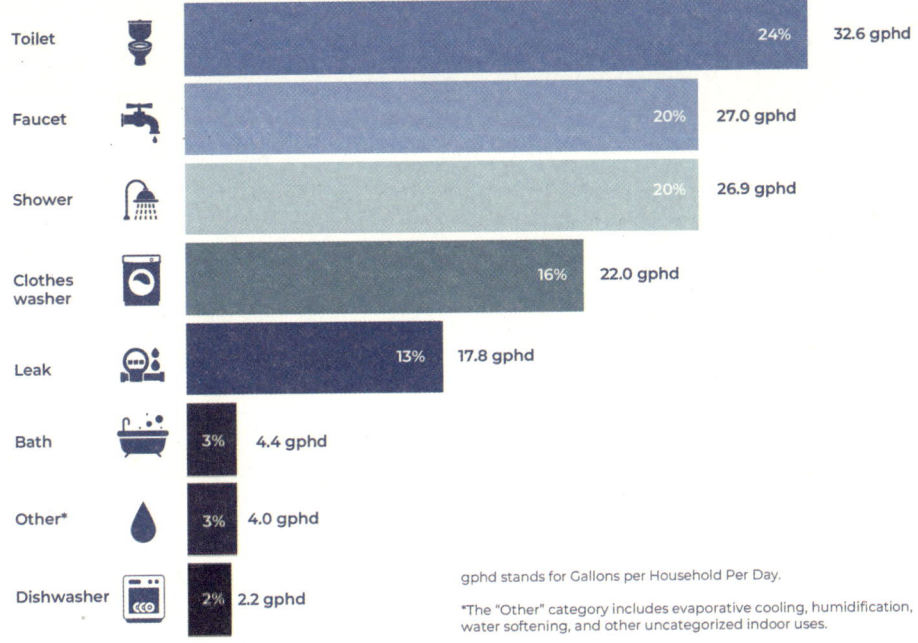

Source: The Water Research Foundation, *Residential End Uses of Water, Version 2*, https://www.circleofblue.org/wp-content/uploads/2016/04/WRF_REU2016.pdf.

So before applying either principle, check to see if your pipes are leaking anywhere. That 13 percent of water usage coming from pipe leaks can be one of the easier ways to reduce impact (assuming you have a good plumber on speed dial). Once you've done that, can you reduce the intensity of your water usage (e.g., using the "little flush" button on your toilet if it has one, taking shorter showers)? Following that, can you reduce the frequency (e.g., showering every other day, washing clothes only when you have a full load, keeping the faucet off when not in use)?

And if you find a good opportunity to replace an old, inefficient appliance with a newer, more energy- or water-efficient one, take it! Because the emissions associated with making things are generally so high, try not to artificially shorten the lifespan of your existing appliances to get more energy-efficient ones. Wait until there's a clear reason to upgrade.

Discussion Points

1) Does the relative energy or water usage of any of the appliances listed in this section surprise you? Why do you think that could be?

2) What could be some lower-impact alternatives to using the appliances listed in this section? What is preventing you and other people from pursuing them?

Closing Part One

I hope that it has become clearer to you how the outsized **negative impact of making things versus using them can provide us a framework** on how to act more sustainably in our daily lives. And for those things that buck that trend (most specifically, houses and cars), I hope the discourse on how to **embrace more efficient utilization of those assets** (rather than immediately replacing them and incurring more manufacturing emissions) can give us concrete levers to pull our lives as close to net-zero impact as possible.

To succinctly wrap up this section, it would ultimately come down to our earlier themes:

1) **Reduce:** If you don't need it, don't buy it.

2) **Reuse:** If you do need it, consider buying secondhand or renting it. (If it must be new, consider greener alternatives.)

3) **Recycle:** If you're done with it, find a new owner for it (if possible), or recycle it.

But ultimately, even looking across all of these levers we have available to us in our personal lives, we are only making decisions as individuals or families. In the next chapter, we'll take a look at how we can pull sustainability into the professional sphere—our jobs, our savings, and our investments—and watch our impact multiply.

Discussion Points

1) What are the top things you could do in your personal life to reduce your impact? What might prevent other people from doing the same? Is there a way to make it easier for them to do so?

2) What do you think are the biggest misconceptions that people might have about their personal emissions? What do you think could help better educate them to focus on the "right things"?

3) Are there things that companies or regulations could do to make it easier for you and other people to understand the relative impact of things? What do you think those might be?

PART TWO: OUR PROFESSIONAL LIVES

Chapter 6: Every Job Can Be a Sustainability Job

This is a rallying cry I've adopted from Drew Wilkinson, one of the founders of Microsoft's employee sustainability community, which is over nine thousand members strong as of the writing of this book.[121] When looking at sustainability careers, it is easy to assume that the number of sustainability jobs is quite limited (i.e., people working in renewable energy, climate science, or nature-based nonprofits) and that the vast majority of work (including investment) has nothing to do with sustainability. Thankfully, this could not be further from the truth!

I field questions almost weekly about how people can pivot their careers toward sustainability. I love these conversations, and one thing I find quite often is that a person's concept of a "sustainability career" is rather narrowly scoped to the above. Oftentimes, people are fixated on the least available but most competitive sustainability roles rather than taking a broader view aligned more closely to their existing work history and skill sets. We can do a much better job as a society of making sustainability a more approachable, understandable career. And first, we need to move beyond the misguided notion that the only way to do sustainability as a career is to have "sustainability" in the title of our roles.

What Are Sustainability Jobs?

Generally, there are three "types" of sustainability jobs. The first type—and one that people think about most—centers on the people who work 100 percent full-time on sustainability in their jobs. They range from climate scientists and carbon accountants to clean-energy procurement experts. They're often embedded within larger organizations or specialized nongovernmental organizations (NGOs) and often require explicit training or education (often graduate-level) in a sustainability-related field.

The second type of job is one at a sustainability-first organization. If you

---

121 Microsoft, *2022 Environmental Sustainability Report*, 2023, https://query.prod.cms.rt.microsoft.com/cms/api/am/binary/RW15mgm.

work for a renewable energy start-up, for example, any position at that company—from sales to finance, operations to marketing, HR to installation—can be a sustainability job, as the services or goods being sold are sustainability products. Without these fundamental roles, these companies would go out of business right away!

The third type of job (and by far the largest) is one that we add sustainability on top of. While there will always be a need for sustainability experts and specialized companies, the reality is that the majority of positive (and negative) impacts within most companies can come from people with "regular" jobs that don't have sustainability in their titles whatsoever. In this chapter, we'll start by spending some time recontextualizing myriad "regular" jobs to show how they were secretly sustainability jobs all along.

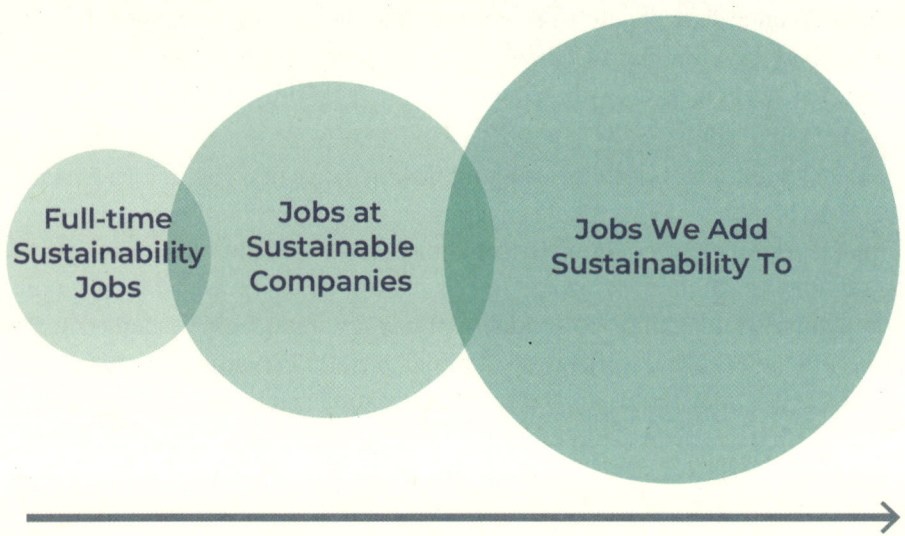

Number of Jobs Available

What Are Your Workplace Emissions?

Understanding that every job can be a sustainability job is a great concept to internalize. For most of us, **the scope and reach of our jobs can yield significantly more impact than the decisions we make in our personal lives**. Though we may not realize it fully, most of us are actively creating additional waste, water usage, and emissions by virtue of the jobs that we do. These emissions come from things like the products we build and sell, the computers and servers that maintain our company websites, the supply chains that transport our goods, and the energy used to power our offices.

WHAT WE CAN DO

To put it in perspective, as mentioned in part 1, the average American person emits 1,400 tons of carbon dioxide over their entire lifetime, or about 18 tons per year. However, while there are no official statistics on this, we can estimate that the average employer emits around twenty times this amount, or around 375 tons of greenhouse gas emissions annually. **If you work, your "cut" of workplace emissions comes out to around 30 tons of carbon dioxide each year, or nearly twice what you emit in your personal life.**

We can estimate this by taking the total gross amount of greenhouse gas emissions produced in America in 2022: around 6.3 billion metric tons, roughly the same as burning down 100 billion trees (yikes!).[122] Then we can estimate how much of this comes from employers using EPA data in its *Inventory of U.S. Greenhouse Gas Emissions and Sinks*.

Let's revisit this information that first appeared on page 10 but look at it in a different way:

- Of the 6.3 billion tons of American emissions, 1.9 billion (30 percent) comes from industry.

- Another 1.8 billion (29 percent) comes from transportation.

  - Out of the 1.8 billion from transportation, about 0.6 billion (9 percent of all emissions) comes from commercial vehicles, flights, pipelines, ships, and rail.

- The next 1.0 billion (16 percent) comes from commercial buildings.

- Another 0.9 billion (15 percent) comes from residential buildings.

- The remaining 0.7 billion (10 percent) comes from agriculture.

By subtracting out residential and personal transportation emissions, you are left with 4.2 billion in emissions, or two-thirds of all American emissions. Divide this by the number of employers in 2022 (11.5 million establishments, per the US Bureau of Labor Statistics[123]), and you get an average emissions per employer of about 375 metric tons of greenhouse gases per year. And if you divide this by the average number of employees per employer (around thirteen), you arrive at an average of around 30 tons of emissions annually per employee in the US.*

---

122 "Sources of Greenhouse Gas Emissions," United States Environmental Protection Agency, last modified June 5, 2024, https://www.epa.gov/ghgemissions/sources-greenhouse-gas-emissions.

123 "Quarterly Census of Employment and Wages: Employment and Wages, Annual Averages 2022," US Bureau of Labor Statistics, last modified November 16, 2023, https://www.bls.gov/cew/publications/employment-and-wages-annual-averages/2022/home.htm.

*I am aware that this leaves out some complexities such as Scope 3 emissions (i.e., emissions from American companies that happen in other countries, such as manufacturing products abroad). These likely mean your emissions from work are even higher than estimated above. But remember the aim of this book: to align with the right comparative orders of magnitude, not precision!

If you ever want to estimate this for yourself and your place of work for a greater level of specificity, divide your company's total emissions (if known) by the number of employees. My employer, Microsoft, emits around 70 tons of emissions per employee as of 2023.[124] If you'd like a sense of *where* your company's emissions come from, check your company's annual sustainability report, if it has one. These documents often break down the company's total emissions into different categories and sources. This can help inform your own "workplace pie chart," similar to the "personal life" emissions pie chart from part 1. For example, the following figure shows my "work emissions pie chart" from Microsoft.

---

[124] Microsoft, *How Can We Advance Sustainability? 2024 Environmental Sustainability Report: Data Fact Sheet*, 2024, https://query.prod.cms.rt.microsoft.com/cms/api/am/binary/RW1lmju.

WHAT WE CAN DO

Scope 3 Emissions Categories (in %)

● Purchased Goods & Services	36.2
● Capital Goods	38.2
● Fuel-and Energy-Related Activities (Market-Based)	3.4
● Upstream Transportation	2.0
● Waste	0.1
● Business Travel	0.8
● Employee Commuting	1.2
● Downstream Transportation	0.5
● Use of Sold Products	14.1
● End of Life of Sold Products	0.1
● Downstream Leased Assets	0.1

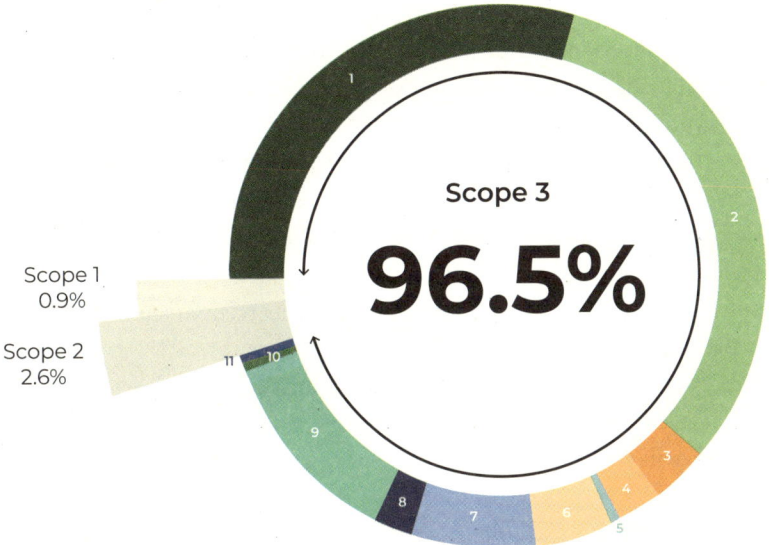

*Numbers may not add to 100% due to rounding.

Source: Microsoft, *2023 Environmental Sustainability Report*, query.prod.cms.rt.microsoft.com/cms/api/am/binary/RW1lMjE

How Are Workplace Emissions Calculated?

To read these sorts of charts, it helps to know about "scopes" of emissions, as this is the language of climate reporting in corporate America:

1) **Scope 1** emissions most often refer to on-site emissions from the burning of fuel, such as driving a company shuttle or running a backup generator.

2) **Scope 2** emissions refer to the emissions generated by the use of electricity on company property, such as to power lighting and AC or run on-site computer servers.

3) **Scope 3** emissions refer to the emissions generated by others on behalf of the company, both the company's suppliers and customers of the products.

Scope 3 is particularly important to measure, as it doesn't let companies "get away" with outsourcing their emissions. Everyone a company does business with will contribute to a company's Scope 3 emissions.

While Microsoft uses a lot of electricity to run its data centers, it is also one of the largest purchasers of renewable energy in the world, helping its emissions from using electricity (Scope 2) stay relatively small. Instead, most of Microsoft's emissions today come from its suppliers and customers (Scope 3). As shown in my work emissions pie chart, this is primarily driven by "Purchased Goods & Services" from the company's suppliers (e.g., those who manufacture its hardware) and by "Capital Goods" (e.g., the construction of data centers around the world). Another major source of emissions from Microsoft comes from "Use of Sold Products," which measures the electricity usage of customers using its products in their homes, like playing Xbox.

Of course, each employer will have its own emissions pie chart, and some jobs will be *materially* more disruptive than others. Someone working on an oil rig, for example, will probably generate far more emissions in their work than someone working as an accountant for a firm. But on average, we can assume that the emissions and waste made at work will generally exceed those of our lives outside it. And depending on the kind of job you have and the size of the company, you could have from a dozen coworkers to hundreds of thousands—not to mention hundreds, to hundreds of millions, of customers. Any impact you can effect at your place of work has the potential to scale impact much further than what you might accomplish alone.

Even small changes at this scale can make a difference. For example, take the hit video game *Fortnite*, which boasts over 200 *million* active players monthly at the time of writing. Each player is using energy to play this game on their devices, and *Fortnite* itself needs to rely on data centers and offices to run the game. And while it might not seem immediately apparent, there are ways you can change software to use energy more efficiently! In partnership with Microsoft and its Xbox Sustainability Toolkit, *Fortnite* was able to identify wasted energy from inactive players and in non-gameplay moments and make minor tweaks to lower resolution and frame rate for those scenarios in ways that were imperceptible to players. These minor changes alone are estimated to have reduced the amount of energy used by *Fortnite* players by the equivalent of fourteen wind turbines![125]

---

125 Unreal Engine/Epic Games, *Reducing* Fortnite*'s Power Consumption*, 2023, https://cdn2.unrealengine.com/reducing-fortnites-power-consumption-layout-v03-ffedbeb1adeb.pdf.

That being said, it can be really hard at first glance to understand how an average job can help influence a place of work to become more sustainable. While there will be some roles with outsized impact potential (like dedicated sustainability leaders or CEOs), there are opportunities for *anyone* to contribute, no matter your role. (And if you're a student at an institution of learning, whenever you see a mention of "your workplace," replace it in your mind with "your school" or your "soon-to-be workplace.")

In the pages that follow, I'll give four examples of roles and companies to provide some clarity into the power we have at our jobs, no matter their type. But first, it's helpful to step back and consider this: If we have the choice, how can we pick a more "sustainable" place of work from the get-go?

Discussion Points

1) Do you feel like your current job could be a sustainability job? Why or why not? In what ways is it already a sustainability job?

2) What is your current company or organization's emissions pie chart? Does any of it surprise you? Where do you see opportunities to influence those emissions?

3) What do you feel could be preventing you from bringing sustainability into your career?

4) What role do you think corporations and workplaces should play in sustainability in comparison to consumers or governments?

Chapter 7: Where You Work

To start, it may be worth reiterating one of the key points from part 1: the physical "where"—in this case, where you work.

Your Commute

A notable source of emissions for many companies (and their employees) comes from the commute that their employees take to get to work every day. In fact, they are required to report on these emissions as a part of their overall emissions profile. (Using terminology from the Greenhouse Gas Protocol, these would be considered Scope 3 emissions, since they're incurred by others on behalf of the company.) At Microsoft, this amounted to 187,000 tons of emissions in 2023—or nearly 500 *million* miles driven by employees![126]

When you consider your next job (or your next move!), one of the most reliable pieces of impact you can have at your job is to live closer to where you work—or to pick a workplace that's closer to home. If you find a job that allows for hybrid work, that's even better, as it cuts down the number of commutes you have to make in the first place. And if you have the option, commuting by carpooling, public transit, or your own two legs (either walking or biking) can eliminate even more emissions. And if you're able to find a fully remote job, a 2023 study made in partnership with Microsoft found that switching from being fully on-site to fully remote could reduce your work-related emissions by up to 58 percent![127]

In short, **using your computer and doing virtual calls (even on video) is far less emissions-intensive than driving**, making remote or hybrid work an attractive way to reduce emissions in your work. In that same vein, skipping business travel in favor of virtual gatherings has an even stronger impact!

---

[126] Microsoft, *2023 Environmental Sustainability Report,* query.prod.cms.rt.microsoft.com/cms/api/am/binary/RW1lMjE

[127] Yanqiu Tao, Longqi Yang, Sonia Jaffe, and Fengqi You, "Climate Mitigation Potentials of Teleworking Are Sensitive to Changes in Lifestyle and Workplace Rather Than ICT Usage," *PNAS* 120, no. 39 (2023), https://doi.org/10.1073/pnas.2304099120.

Your Employer

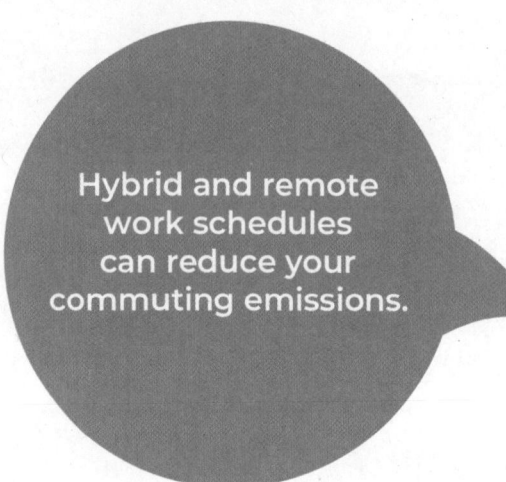

Hybrid and remote work schedules can reduce your commuting emissions.

When picking "where" to work, this includes the company itself in terms of its values and what it stands for. As mentioned above, the more your employer is sustainability-oriented, the more your everyday role can contribute to making an impact on the planet. However, it is often hard to suss out whether a company is genuinely committed to sustainability from the outside.

One of the quickest and simplest ways to assess this can sometimes be a glorified "vibe check." Does the company *seem* like they are committed to sustainable causes and operations? Many companies today fail this vibe check, most often by not talking about sustainability at all. But for the ones that do, there are thankfully a few ways to gauge how serious they are. Generally, though, if an organization is talking about sustainability as a part of their mission, you're already off to a great start.

First, is sustainability a clear part of their brand or mission? Companies like Patagonia or Seventh Generation put sustainability front and center when making and marketing their products. Others may talk about sustainability as important but haven't yet made the leap toward making it a primary identity of the company.

Even if it's not a part of a company's brand, that doesn't mean a company isn't necessarily committed to making an impact. It's often worth checking to see what a company's sustainability commitments are. Many large companies today have made major commitments out to 2030 or beyond to drastically cut their emissions, waste, and water usage. Typically, the sooner the goal, the more aggressive the company will have to be in making sustainability a real part of their operations.

However, this has to be backed up with results. Have independent groups ranked a company as having a high environmental, social, and governance (ESG) score? Has a company's emissions or emissions intensity started going down, or "avoided" emissions gone up, in recent years? Has it shown investment in a long-term pathway to decarbonize? Or, for smaller or more local organizations that might not have the capacity yet to be measured in this way, have they made material changes in their stores or products to reduce waste?

WHAT WE CAN DO

Your North Star

As part of choosing the right place of work, it's also crucial to ask yourself what matters most to you. For instance, is your goal to maximize your total positive impact on the planet? Jobs at companies that are some of the worst contributors to waste and emissions can often be ones with some of the greatest *positive* impact potential.

Take the tobacco industry, for example. As mentioned in part 1, per the National Oceanic and Atmospheric Administration, cigarette butts are one of the most commonly found types of ocean litter. Many tobacco companies have begun embracing sustainability goals in the meantime to reduce their impact, creating space for employees to lean into the challenge. Helping transition the tobacco industry away from nonbiodegradable cigarettes would yield immense benefits in reducing ocean plastic waste.

Now, what if your goal is to maximize your impact on your local community? In this case, maybe a job at your local parks and recreation department or a locally focused nonprofit could fit that bill. While the impact on the entire planet might be smaller in comparison, cleaning up your immediate community can be very rewarding, as this impact is something you can interact with every day.

> Working for a sustainability-forward company can amplify your sustainability impact.

And maybe your goal isn't measured in impact on others. Rather, maybe it's most important to you that *you* feel like you can work on sustainability 100 percent of the time at a company whose mission you fully support. For example, working at an environmental start-up might end up yielding almost no positive impact on the planet at all (as is the risk in the start-up world), but it may personally rejuvenate you.

I cannot answer this question for you. But at the minimum, I encourage you to answer this for yourself when looking for a sustainability job: **What is your North Star?** At least on my end, I have made my North Star to maximize my positive impact on the planet. (Otherwise, I wouldn't be writing this book while working in sustainability at one of the world's largest companies.) I get more energy when making a bigger, global impact than a smaller, local impact. However, I fully acknowledge that this may change one day as my career progresses and my stage of life changes.

With all this being said, once you have a job ready to go, how can you drive sustainability into what you do?

Discussion Points

1) When it comes to your career, what is your North Star?

2) Which companies come to mind as sustainable companies? What makes them appear as such? How can you tell whether they actually are?

3) What are some potential downsides to hybrid or remote working that could create additional emissions?

Chapter 8: At Your Work (If You Work a "Normal" Job)

When I claim that every job can be a sustainability job, I really do believe it. However (and very understandably so), I think it's not at all clear for many people how many jobs can have a sustainability component. At a high level, I'd offer up this CORE Framework for things we all can do at our jobs or campuses:

1) **<u>C</u>onnect to Impact:** This principle is really about helping others—whether they're people inside your company or your customers and partners outside of it—understand what the impact is when they engage with your company. Oftentimes, just helping to expose which choice is more sustainable can help drive better behavior or at least build awareness for the future.

2) **<u>O</u>rganize Internally:** This principle is about influencing peers who may be in roles that can have a direct impact on sustainability. Many companies have local chapters or groups of like-minded individuals that often span many types of job roles and seniority. Helping them to understand the role that they can play is a way you can outsize your impact.

3) **<u>R</u>educe Waste:** This third principle is focused on what you can do in your existing job. For the most part, each of us either creates waste at our jobs daily or is attached to decisions that can determine the amount of waste that is created. Oftentimes, reducing waste can also reduce cost—both for you and your company—and so it is often simple and easy to do.

4) **Enact the Change:** This last principle is among the most proactive. It often means looking at your span of influence in your existing job and actively working to build your role into one that can have more influence on sustainable business practice. On paper, it can seem the hardest, but this approach is one of the best ways to build your career into a sustainability career.

Let me share four examples of everyday jobs that, at first glance, have nothing to do with sustainability but can leverage the frameworks laid out in this book ("Reduce, Reuse, Recycle") to transform into jobs that do. I'll connect each one to the categories in CORE.

THE SOFTWARE START-UP SALESPERSON

THE CORPORATE EMPLOYEE

THE FOOD SERVICE WORKER

THE EVENT PLANNER

Connect to Impact: The Software Start-Up Salesperson

In its essence, "Connect to Impact" is about helping people understand the impact of the things that they are already doing. Think about the "Make It, Move It, Use It, Lose It" framework, for example, and how it can contextualize the "Reduce, Reuse, Recycle" hierarchy to unlock new ways of seeing impact. In many cases, this can mean helping people realize that sometimes they have been making more sustainable choices for a long time but have never put those choices in the context of impact!

Let's take the role of a salesperson at a start-up that sells security software. Their job is to sell software solutions to other companies to help them experience fewer hacks and breaches. They're not involved in making decisions on how the product is built, however, and you may be wondering whether a security software company really has that much of an impact on sustainability at all.

It's very fair to start from that assumption. However, it's most often an inaccurate assumption. The beauty of digital solutions can be in their capability to solve problems without needing to make physical things.

In this case, let's say that the salesperson has been talking to prospective customers who have been complaining that they need to buy new machines for all their employees to help reduce the chances of being hacked on older, more vulnerable hardware. Or perhaps they are frustrated that their existing security solution takes up a ton of bandwidth on their network and computers, causing their employees to complain about slow devices.

Assuming that this salesperson is selling a better solution, both of these instances can be places to "Connect to Impact." In the first case, if better security software could let this company use its machines for longer, then you can draw a direct line to the amount of "avoided emissions" and waste by calculating the number of PCs and equipment it no longer needs to buy, then multiplying that by the per-device carbon and waste footprint. If this is a two-hundred-person company, you could be helping it avoid emitting 40 tons of carbon—the equivalent of 50 acres of forest!

To go even further, if their solution runs more efficiently on their hardware, not only can that delay their need to buy new, faster machines, but it can also reduce the amount of energy used to run the security solution, saving the customer money on their energy bills as well as emissions.

In this way, the salesperson can make a big difference by just making this impact visible to the customer and their own employer. Shining a light on an entirely new piece of the value of the product provides potential differentiation to the salesperson to stand out in the field—and it could ultimately feed back into the marketing and product teams. And who knows, maybe the marketing team gets so inspired that they put out an Earth Day promo to plant trees as part of every deal!

Finally, once the salesperson can translate their work to impact, this can help them make more sustainable choices as well. For instance, can they land the deal through a virtual meeting instead of traveling in person? Or take the prospective customer to an eco-friendly restaurant instead of a golf outing?

> **Discussion Points**
>
> 1) Think about your job. Are there things you do or interact with that could enable more sustainable outcomes?
>
> 2) What is preventing you from measuring the impact of your work? Are there high-level or proxy calculations you could do to estimate it?

Organize Internally: The Corporate Employee

At the start of this chapter, I mentioned my friend Drew Wilkinson, who was one of the founders of Microsoft's nearly ten-thousand-person employee sustainability community[128] and now serves as the founder of the Climate Leadership Collective, an environmental consulting firm focused on employee engagement. Initially, however, Drew was an associate paralegal within Microsoft's legal team.

"I came to Microsoft from the nonprofit world, and prior to that, I was in a punk rock band," Drew shared in an interview with me for this book. "This huge shift in my career led me to ask myself on an existential level, 'What am I doing here?'"

Despite his paralegal job having nothing to do with sustainability, Drew retained his passion for the planet and started to ask questions: "Coming from a resource-constrained nonprofit of five people, it blew my mind seeing hundreds of buildings, thousands of employees, and a private shuttle fleet at our headquarters. It led me to ask, 'Could this company use some of its resources for climate change as well?'"

So Drew started poking around, asking two simple questions to his coworkers and leaders:

1) "What is this company doing for sustainability?"

2) "How can I get involved?"

---

128 Drew Wilkinson, "The Critical Role of Employee Sustainability Communities and How to Build One," Microsoft (blog), April 18, 2023, https://techcommunity.microsoft.com/t5/green-tech-blog/the-critical-role-of-employee-sustainability-communities-and-how/ba-p/3795108.

Initially, he found an email group of employees who was passionate about the environment, but most of the discourse was complaining about the status quo, from seeing gas-powered leaf blowers to finding food and plastic waste in the dining halls. But as Drew recalled, "nobody was willing to offer up thoughtful alternatives to these problems. Nobody was willing to challenge the status quo or talk to the decision-makers and find out why we had made these decisions."

Rather than being content with just complaining about the problem, Drew sought out like-minded, action-oriented peers who were willing to advocate for change, no matter their job or background. Luckily, he found a fellow action-oriented peer who had independently asked the company's facilities team about getting access to the company's waste statistics. "I've found that waste is the 'gateway drug' of getting involved in sustainability at work," Drew said, laughing. "As you move through your work environment, waste is the most noticeable part of that flow and a visual reminder that we can do better—and a great place to start advocating for change and getting people's buy-in."

With data on the company's trash in hand, Drew and his peer requested an audience with the people in charge of waste within the facilities team, then presented a list of near- and long-term recommendations based on the types of waste generated. Six months later, the decision-makers called them back in and showed them that they had instituted a number of zero-waste initiatives based on their recommendations.

"This was the light bulb moment for me," Drew recalled. "If two random employees without sustainability in their job titles—or really any power at all—can get a company the size of Microsoft to change operational practices to become more sustainable, what if there were thousands of us advocating for change?"

So, with a résumé that would otherwise "never get him a sustainability job," Drew set out to see if he could bring together like-minded peers who were willing to advocate for change at scale.

To grow the community, he first sought executive sponsorship. "Organizing around sustainability shouldn't have to be a secret, or something that employees do in the shadows," Drew guided. "If you do it in the open—and do it in a collaborative way to how the company works—you can get sponsorship that can unlock resources."

This executive sponsorship helped lead to the sustainability community being recognized as a formal "employee community" alongside other recognized employee groups, like developer communities. It also helped Drew find global teams who had independently started their own local sustainability groups but had not been connected together. Pulling the community under one "big tent" with local chapters helped it grow while encouraging local action. It also sourced peers who were willing to volunteer their time to help maintain the community behind the scenes.

"Ultimately, it was important to socialize that ALL employees deserve to be a stakeholder and to have a seat at the table," Drew said. "We wanted to change the narrative of who gets to work on sustainability. It should be part of everybody's job."

This helped to amplify employee sustainability advocacy in a scalable way. "In many ways," Drew continued, "employees have an incredibly privileged role in these discussions because we're already inside the machine. We can have private conversations with our leaders that are more frank and difficult than what a company might be willing to discuss publicly."

When Microsoft announced its landmark "Carbon Negative" climate commitment in 2020, the sustainability community membership exploded. "It was amazing to think that our voices as employees helped contribute—even if a little bit—to the company's decision to level up its sustainability ambitions," Drew said, beaming. "Once we had a top-down mandate as a company on sustainability, our community attracted people that would never have thought of getting involved before."

And while Drew has since moved on to build a new organization, Microsoft's community continues to thrive, with nearly ten thousand members as of 2022. "It's kind of funny," he said. "Even with all of my work with the sustainability community at Microsoft, until I started the Climate Leadership Collective, *I never actually had a job fully dedicated to sustainability.* I had to go and invent one based on the groundwork laid in my normal corporate job."

So with this as an excellent example of "Organize Internally" to build sustainability into his career, Drew left me with a set of advice for others who may be interested in doing something similar:

- **Just Get Started:** You don't need to be an expert in sustainability or community organizing to get started. You just need to start and be unafraid to try rather than wait for someone else to make it happen. If you build it, they will come!

- **Look Around:** Assess whether there are other peers doing similar work in pockets. If there are, rather than duplicate effort and divide, try to plug in, volunteer, and help to improve, grow, and amplify their work—or bring them into your own efforts.

- **Leverage Your Skills:** Look at the intersection of your current job with sustainability and then reimagine how it could be done with a reduced environmental impact. If you figure something out, it could help EVERY person across all companies and industries.

- **Get Support:** Go to your leaders and ask them, "What role do employees play in sustainability at this company?" If they don't have an answer, that's a great invitation to help define it together.

As a final thought, Drew encouraged us to think of getting into sustainability not as a light switch but as a dial you turn over time to build sustainability into your role. "Don't let sustainability being in or out of your title be a psychological barrier to making the work happen! Leaning into sustainability from where you already sit will give you the network and experience to eventually get you involved in climate, even if it takes years and comes in a very unexpected way, like it did for me."

Learn more about Drew Wilkinson at his website: https://www.drewwilkinson.earth/.

Discussion Points

1) Think about your company. What could be the "gateway drug" that could galvanize your peers to organize on sustainability? Who may be willing to sponsor this work?

2) Reflect on your current role. How could you leverage the skills you have in your job to help support a sustainability community at work?

Reduce Waste: The Food Service Industry Worker

This category is one that is near to my heart, since my first job was at the fast-casual sandwich chain Jimmy John's. As a high schooler, I was a minimum-wage employee in Wisconsin who was not yet qualified to be a delivery driver, so I would work front-of-house (e.g., making sandwiches, operating the register) and back-of-house (e.g., cleaning the kitchen, prepping the ingredients).

As a minimum-wage worker, my formal job accountabilities were very clear (and nonnegotiable). I was supposed to make sandwiches. I wasn't there to swap in more sustainable cookie packaging or bring in new, more carbon-friendly sandwich options. Many of us are in roles like this that are tightly defined, from food service and baristas to office administrators, retail employees, and construction workers.

As important as it is to understand what our contracts dictate, we must also understand where there is flexibility to make changes around the edges. In Jimmy John's case, I realized that the sandwich shop was a great place to reduce waste around our shop's bread policy at the time.

Jimmy John's bakes all of its bread in-house, and the amount of bread baked is done in anticipation of how many sandwiches we expect to make in a given day. In the food industry, it is always safer to overestimate rather than underestimate the amount of food you need. The philosophy behind this is that it's better to have some food waste than run out of food with customers left to serve.

At the time, all of the store's excess bread was saved overnight and then—because it had lost some of its day-of fresh taste—was not used for sandwiches the next day. Instead, it was sold as "day-olds" for dirt cheap (which was a bargain, as the bread was still delicious). This, by itself, was a great way for the store to recoup some of the cost instead of throwing all the unused bread away the day it was baked. But any day-old loaves not sold that day ended up being thrown in the trash.

Because there was no rule against it, during every closing shift I worked, I would volunteer to take home all the day-old loaves that would otherwise be thrown out. My shift manager took no issues with this. For the store, the result was the same; the old bread would get removed from the store inventory. For me, however, this meant a chance to repurpose the bread that otherwise would have gone to waste. It turned into many sandwiches and slices of garlic bread at home, or even croutons or bread pudding (a great thing to make with stale bread!). For other types of food waste, there are many food banks and shelters that would happily accept the excess food to feed the hungry.

Many of our workplaces will throw unused things out all the time to save themselves the hassle of having to deal with them. Volunteering to reduce this waste is often (at worst) neutral to, and (at best) helpful for, the stores. And in some cases, it may actually influence the store to build a formal program to transfer its unused goods to nonprofits or other groups in need.

> **Discussion Points**
>
> 1) What kinds of waste do you typically see in your current job? What is causing that waste?
>
> 2) Are there ways that your company could change its policies to reduce this waste? Are there ways that you can reduce the waste yourself?

Enact the Change: The Event Planner

The final category to bring sustainability into your work is to Enact the Change needed to make your role a more explicit sustainability role. While this is often the hardest category, at many companies, there is a lot of "unclaimed territory" when it comes to sustainable impact, which creates space for many individuals to expand their role to be more impact-focused.

I'll use the example of an event planner. Whether in a corporate (e.g., conferences) or personal (e.g., weddings) environment, event planners are responsible for making sure that the events that are being hosted go off without a hitch. They are often in service of the customer or company's overall theme but are tapped to fill in the details of how the event can come to fruition seamlessly.

Unless the event owner is VERY particular and dictates how every minute detail needs to be, chances are that the event planner has a decent amount of leeway in what kind of things are procured to support the event, how the attendees get around the event, and even some of the extracurricular activities available or the types of food on the menu! Each of those things is an opportunity to "Enact the Change" and bring more sustainable options to bear.

For many events, the biggest source of emissions tends to be the transportation of all the people (and goods) to the event. A savvy event planner may recommend or directly organize shared transit to get to the event (e.g., shuttle buses) rather than have people drive independently, especially if attendees are mostly staying in the same hotel or neighborhood. They may also think of using some of their budget to offset some of the travel emissions (especially flights), perhaps by paying to plant a few trees for every flight taken. And if the event planner has any control over the venue location, could they find one that is easier to get to from public transit or is more centrally located?

Within the event agenda, there could also be really fun ways to engage the attendees on more sustainability-themed events. For example, instead of a golf outing, could the event host a tree or garden planting or another nature-based warm-up event? Could there be raffles for items where the proceeds go to an environmental nonprofit? Could the event partner with a cause that encourages the attendees in their own time to go to a website to make an impact or learn more?

Once at the event, there is often a LOT of single-use materials that go to waste directly afterward: tablecloths, signage, drink containers, and even goodie bags or handouts. However, the event planner can look to tackle waste here by virtue of whom they order from and what they order. Are there more sustainable alternatives that are being sold by vendors of these materials (e.g., ones that aren't made of single-use plastics)? When thinking about what's in the goodie bag, for example, this often means trying to avoid water bottles and plastic chip or cookie bags (things I seem to receive every time I get a goodie bag at an event). Even better, if the event planner wants to cut down on costs and help the environment, can they use as many reusable materials as possible? And at a minimum, can recycling and composting bins be readily available across the venue?

For the things consumed at an event, the event planner has an opportunity as well. Rather than over-ordering food, for example, can they reduce food waste by keeping the ordered food amount closer to the expected attendance? Can they aim to serve smaller portion sizes, which are more likely to get completely eaten? Also, can the choice of hors d'oeuvres or main meals be more locally sourced? And can they avoid red meat in favor of lower-impact meals, like chicken or vegetarian options?

Energy is consumed at an event too—from lighting and AC to wedding DJs—and is a potential place to "Enact the Change." While it's highly unlikely that a venue can install solar panels in time for an event, there are often ways to purchase cleaner energy from the electric grid, which the event planner could include in their contract with their event space. They can also look to places to drive more energy efficiency, whether in the amount and types of lighting used or the amount of AC used. (Ever been to an event where you're unnecessarily freezing the whole time? That excess AC wastes a lot of energy.)

If the event planner so chooses, this focus on sustainable events can easily become a focus area—and perhaps a competitive differentiator—to their business! It's not just theoretical: Here in Minnesota, there is a wedding venue—Paikka—which offers a 10 percent discount for couples that sign up for their Sustainability Program for weddings.[129] And over many events, event planners

---

129 "Sustainability at Paikka", Paikka, accessed July 2024, https://www.paikkamn.com/

like these could start to calculate just how much single-use plastic didn't go to landfills or how many emissions were avoided by using shared transit.

To close this last example, let me share an awesome real-life expression of this in combination between event planners and event owners—in this case, music events. The megaband Coldplay has started to release sustainability reports for its concerts and talk in detail about how every element of the event is reconsidered to drive for more sustainable outcomes. The results? Coldplay's 2023 "Music of the Spheres" tour produced 47 percent fewer greenhouse gas emissions than its last stadium tour ending in 2017, planted five million trees (one for every concert goer!), produced an 86 percent return rate on LED wristbands (which, in prior years, would have been used once and discarded), and diverted 66 percent of all tour waste from landfills.[130] This is a great example of "Enact the Change" in a space that historically has barely ever considered the impact of its operations.

Discussion Points

1) Reflect on your current place of work. What could be some "unclaimed territory" on sustainability work that you could explore? What barriers might prevent you from doing so?

2) Do you know of anyone else who's embedding sustainability into their work? What approaches have they taken to be successful?

Can CORE Be Applied to You and Your Work?

This is just one small fraction of the types of roles out there. But for the most part, all of them can leverage at least one of these four principles of the CORE Framework to help make a more positive impact at work. And as the world becomes more conscious of the role each of us can play, more and more supporting organizations and educational materials will start to emerge, such as the Green Software Foundation, which aims to help every software developer learn how to write more sustainable code.

Do you have a job or role that you'd like to layer sustainability on top of

---

[130] "Tour Emissions Update," Coldplay.com, June 3, 2024, https://www.coldplay.com/emissions-update/.

but don't quite know how? The beauty of modern technology is that our conversation doesn't have to end in the pages of this book. We are building a running list of examples on my website, so feel free to add to the discussion at www.CharlieSellars.com.

Chapter 9: At Your Work (If You Work a "Sustainability" Job)

Before shifting gears, I want to dedicate some time to shine some light on the current state of "sustainability" jobs in the world. For some of us, we may not be content to have sustainability as just a portion of our careers. I know I specifically sought out a way to make sustainability my full-time career. To start, I will share a bit of how I got into sustainability as a career with no formal background in it by "making sustainability my job," as I described in the paragraphs above and aligned in particular to the E of the CORE Framework. Then I'll look to more robust sources of data to share with you what the broader landscape looks like on this topic.

How I Got My Own Sustainability Job

When I graduated from college, I did so with a degree in physics. I had initially thought I would become a physicist or an engineer and never once took a class on sustainability while at school. However, not being smart enough to be either a physicist or an engineer, I ended up taking a job at a company focused on providing advisory services to other large companies. (The irony doesn't escape me). Initially, my role as a research analyst was scoped specifically to write long-form essays for clients in IT in almost a private "think tank" capacity. These topics had nothing to do with sustainability; rather, they focused on technological and business topics like "The Internet of Things," "The Future of Work," and the like. And while I did attempt a few times to encourage my managers to let us write on sustainability, I was told gently that this was not something our clients would pay us to research.

Unbeknownst to me, I was building a set of skills that would translate quite well into a career in sustainability. The research we were accountable for writing required me to take complex, future-looking topics and distill them into simple yet strategic messages for companies. When I had to present these papers to our clients in person, this forced me to build a background in how to engage with company leaders and "speak their language." In some ways, my first job made me into a translator between the subject matter experts deep in their fields and the business leaders who needed to know only the most important details to make their next decision. This skill set is still one I use every day in my current sustainability role.

Four years into my first job, I realized the emergence of a new North Star for myself and my career. I wanted to use my career to maximize the positive impact I could have on the planet, and the field of IT had not yet evolved enough to be the right place for me to do so at that time. So I talked to my boss about changing roles within the company to allow for more of a focus on sustainability, and we found a role for me in the Supply Chain Advisory portion. Contrary to IT, Supply Chain was already waking up to the need to focus more on sustainability. Supply chain owns the "Make It" and "Move It" parts of their companies which, as we've learned, tend to represent the most impact on the planet. Supply Chain had the levers to pull to make a difference.

In this new role, I had more leeway to determine what our clients would come to meet and discuss. So, because nobody told me I *couldn't*, I started making sustainability a part of my role and my personal brand. Whenever we would assemble our clients for a conference or meeting, I would always make sure to have at least some content on sustainability, often asking one of our clients to help deliver the content or drive the conversation.

But this wasn't just me pushing an agenda. Our clients *wanted* to talk about these topics and rated these sessions highly. Sustainability was a value-add to the company; it just hadn't known it yet.

After three years of building sustainability into my job, an opportunity arose at one of our client companies, Microsoft. It had realized the value in hiring a program manager to coordinate sustainability efforts across its Windows and Devices business group so that it could become a more strategic priority across the product line and its supply chain. And because I had spent three years building my personal brand on sustainability and organizing sustainability-themed content for my clients, my name rose to the top when they were considering who to ask for the role.

For this role, it was less important for me to have a formal background in sustainability. As it was explained to me, Microsoft was more concerned about finding someone who it felt could act as the "translator" of sustainability into a strategic business initiative with executives and as a "builder" and program manager, which had been my role in growing my old company's Supply Chain Advisory programs. And—especially important for sustainability—it wanted someone who would be a champion for the cause and build a culture around it.

Since joining Microsoft, I have learned from all of my excellent peers with more traditional sustainability backgrounds to build my knowledge base, as well as from my business partners on how the business actually operates. My work's charter, both for my old role in the Windows and Devices business and my current role in the Cloud business, is (most simply put) to orchestrate all parts of our business group to ensure we have embedded sustainability into how we operate and to make sure we have a strategy to maximize our positive impact on the planet and for our customers.

As to why I firmly believe that every job can be a sustainability job, many of the people who actually *make* the sustainability impact in our team are people with specialized non-sustainability backgrounds. They are the industrial designers, the engineering managers, the sourcing managers, the logistics experts, the software developers, the energy gurus, the marketeers, and many more who are *really good* at their regular jobs and can layer sustainability on top of those jobs. I'd much rather that an expert at designing products learn how to make a more sustainable product rather than try to teach myself how to build products from a background in sustainability! (Remember, I was not smart enough to become an engineer out of college.)

> Your career is one of your biggest opportunities to create a positive impact.

My story is just one of many, but hopefully it can help show some common themes that I used to switch into the field (e.g., building key business skills, building sustainability as a personal brand, making sustainability a part of my job, "Enacting the Change"). There are also many more traditional sustainability jobs for people with actual training and backgrounds in sustainability, like my excellent peers in full-time sustainability roles I work with every day.

The Skills That Can Enable You to Move into a Sustainability Job

One benefit of working at Microsoft is the access to job trends data that it gathers through LinkedIn, which it purchased in 2016. In its 2022 report titled *Closing the Sustainability Skills Gap: Helping Businesses Move from Pledges to Progress*, it shared that the International Labour Organization estimated 18 million net new jobs will be created by 2030 as a result of attempting to meet the goals of the Paris Agreement, a landmark agreement across multiple countries to try to limit global warming to 1.5 degrees Celsius (2.7 degrees Fahrenheit). But while green jobs grew at an annual rate of 8 percent between 2015 and 2022, the talent pool only grew by 6 percent.[131]

Today, of this fast-growing set of roles, the majority (57 percent) of professionals currently in sustainability jobs do not formally have a degree in sustainability, while 43 percent do. And though it's still not as popular as other fast-growing majors, environmental studies is nevertheless growing at a significant rate among undergraduates as well as within many graduate programs. Whether these degrees are growing fast enough to fill the coming expansion of full-time sustainability jobs remains to be seen. Irrespective of educational background, the top five skills to enable a dedicated sustainability career are shown below:

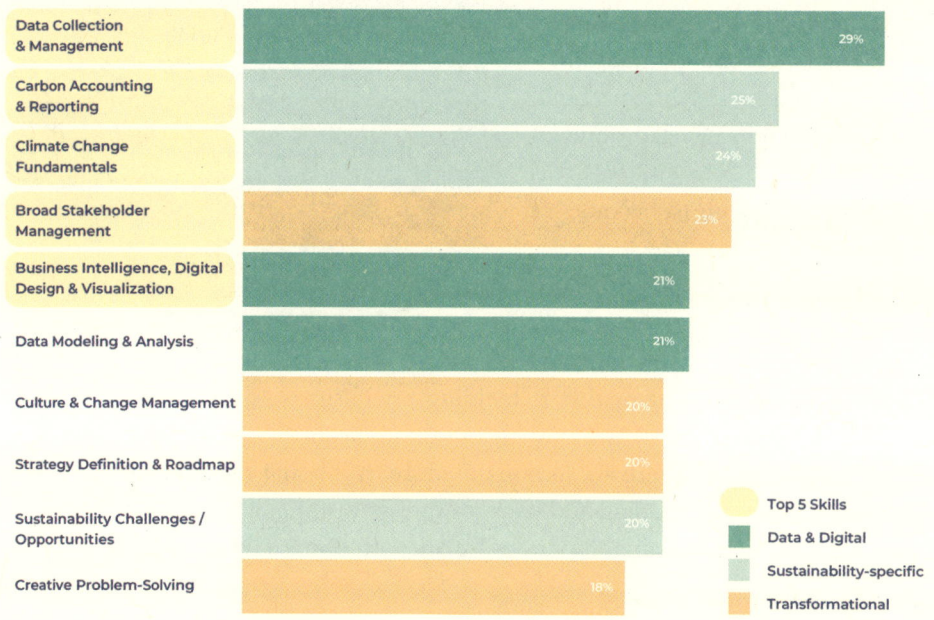

Source: Microsoft, *Closing the Sustainability Skills Gap: Helping Businesses Move from Pledges to Progress*, https://query.prod.cms.rt.microsoft.com/cms/api/am/binary/RE5bhuF.

The first skill, data collection and management, refers to roles that can translate action to impact in a consistent, repeatable way. This is a skill set that already exists for many jobs, hence this age-old adage repeated in team meetings: "You can't manage what you can't measure." What's new here is how sustainability data can layer on top of existing processes or data collection that

---

[131] Microsoft, *Closing the Sustainability Skills Gap: Helping Businesses Move from Pledges to Progress*, November 2, 2022, https://query.prod.cms.rt.microsoft.com/cms/api/am/binary/RE5bhuF.

are already in use by the business. Many full-time sustainability roles in this world can be a hybrid of sustainability and business skills, like being in charge of recording and managing the emissions from your suppliers as part of your supplier-engagement strategy.

The second skill, carbon accounting and reporting, describes the core responsibility for careers in more formalized sustainability reporting. Many companies currently report their emissions, water, and waste usage, either voluntarily or to meet regulatory requirements. These roles can be more technical (e.g., conducting a lifecycle assessment for a product, which sometimes requires a graduate degree and training to excel at) or more lyrical (e.g., pulling together a company's annual sustainability report). Reporting roles often benefit from a background in compliance, given the complexity of meeting reporting standards that exist and the need for deep subject matter expertise to adhere to them.

The third skill, climate change fundamentals, is akin to the traditional "climate scientist" role but is within a company. These are the types of roles that require a deep understanding of some element of sustainability and are often tapped to engineer or advocate for cutting-edge solutions to existing problems or to define and procure sustainability solutions or offsets in the market. This could also include roles that procure renewable energy to power your facilities. These, too, are often roles that benefit from graduate degrees.

The fourth skill, broader stakeholder management, is a big part of what my type of role looks like. It applies mostly to program and project managers for sustainability. These are often the roles that need to coordinate across many different stakeholders (whether they're sustainability experts or not) within a company to align with a plan or strategy to enact the change needed to enable a more sustainable outcome.

The final skill, which I'll combine and summarize as data visualization and analysis, is for the data scientists and data analysts of the world. As sustainability data becomes more visible for the first time in many companies, it becomes critical to visualize current-day performance and to model complex forecasts for years into the future to understand if and how companies can hit their big 2030, 2040, or 2050 climate targets.

What Kind of Sustainability Career Do You Want?

While there are many more skills that can help build a successful career in sustainability, having this mix of sustainability expertise and data and business knowledge can create a powerful resume. But more important than these skills is the fundamental question that you must ask yourself: **What kind of sustainability work do you want to do?**

Just wanting to "do sustainability" full-time is often not specific enough and doesn't help you build your skills in a way that can make you an attractive candidate. Do you want to focus on renewable energy procurement? Carbon accounting? Sustainability data science? Sustainability product marketing? Or do you not care what role as long as you're at a sustainability company, in which case you could work in any type of role?

This is an answer that I cannot help you with and is a journey where you must lead yourself. Whatever you choose as a career should be something that gives you the energy to make an impact. And once you figure out what specific "flavor" of sustainability you want to do, what is most important is to look at the specific set of skills being asked for in the job posting you may be looking at. Generally, if you are looking for inspiration and want to find an open job in sustainability, using key words like *sustainability*, *environmental*, *waste*, *energy*, *water*, *carbon*, *ESG*, and *climate* can often help you find the right job postings. Consider also searching for work on climate-specific job boards; there are many emerging resources that can tailor your search.

Discussion Points

1) *Sustainability* can mean many things. What kind of sustainability work would provide you the most fulfillment in your career?

2) What are some near-, medium-, and long-term actions you could take to build your career into one of sustainability?

3) What is your "personal brand"? What would it take to make sustainability a part of that brand?

Chapter 10: But What Does Corporate Sustainability Look Like?

Before closing this conversation on employment, I'd love to take a moment to paint a picture of what corporate sustainability *actually* looks like, as well as what I've seen determine more successful approaches to achieving it. For this example, I'll talk specifically about what it can look like in large Fortune 500 companies, where I have the most experience.

What's important to understand about large corporations is that they are not monoliths. Oftentimes, they are engaged in multiple businesses at once, each with their own sets of products, cultures, motivators, and customers. Take Unilever, for example, which makes Axe body spray, Ben & Jerry's ice cream, Dove soap, Hellmann's mayonnaise, Lipton tea, Q-tips®, and dozens more products across three different segments: food and drink, home care, and beauty/well-being/personal care. Similarly, Microsoft has a number of different segments it competes in: PCs and devices (e.g., Surface devices, the Windows operating system), gaming (e.g., Xbox, Xbox Game Studios), productivity (e.g., Excel, PowerPoint, Word, Outlook), communications and messaging (e.g., Teams), internet search (e.g., Bing), social media (e.g., LinkedIn), cloud services (e.g., Azure), and more, including the rise of artificial intelligence (AI) services as a horizontal enabler across each.

A good allegory is to look at how the US is organized. At the top level, you have the federal government. This top level provides services for the whole country, like defense, the Federal Reserve bank, interstate policy and regulation of interstate commerce, international relations, and plenty more. In a large corporation, these "corporate-wide" teams are the ones providing centralized business services, like IT services, payroll, company policy and commitments, corporate reporting, and legal services. Ultimately, they are led by the CEO (or, literally, the "president" in some companies), and they're ultimately responsible for the overall strategy and well-being of the entire company.

Next up are the states. Other than what the federal government mandates of them, they are given leeway on how to manage themselves, such as with gas and sales taxes, maintaining public schools, regulating guns and alcohol sales, and more. These states are like business groups; just as Texas and Maine have two different sets of constituents with two different sets of priorities, business groups have their own priorities that can radically differ. For example, while Microsoft's Cloud team may see Amazon as a competitor because of its Amazon Web Services offering, Microsoft's Devices and Gaming teams may see it as a partner because of its e-commerce marketplace, where people can buy Microsoft's PCs and games. Each of these business groups will have its own president and be responsible for strategy across its business group and its linkage to the corporate strategy.

After that are the cities and municipalities, which control more localized decisions like zoning restrictions, housing ordinances, and commerce incentives. A business group's "cities" are like its products, the core engines of the business groups that sit within the broader corporation. In the Unilever example, the team that manages Ben & Jerry's is responsible for deciding which flavors to get into as well as developing new flavors and combinations. Ideally, they are building off the infrastructure that the state (business group) and federal (corporate team) levels provide.

Also within cities are the countless other mechanisms that make up a city, like school boards, homeowners associations, and parks and recreation departments. In a company, these can correlate to the individual functions that are executed to deliver the product's success. This could be the supply chain team, the packaging team, the marketing team, the product development teams, the sales teams, and plenty more.

At a high level, this is a fun way to think about how a large company is structured. In reality, each company is different depending on the business it is in. Oftentimes, you'll see a company decide to "elevate" a function (like supply chain) so that it can leverage the same function to deliver multiple products at the same time. And sometimes, the business group and the product are the same, so the structure is more compact.

Here's where the fun part comes in: In a company that has invested appropriately in sustainability, *each of those levels requires people working on sustainability* to make it happen. This can lead to hundreds—if not thousands!—of part- or full-time sustainability opportunities at large companies that have demonstrated commitment to sustainability. Here's an example:

- **The Corporate Sustainability Team** needs to coordinate sustainability across all of the business groups. It sets company-wide sustainability policy (internally as well as advocate externally); makes company-wide commitments, measurements, and disclosures (e.g., annual sustainability reports); and is accountable for bringing each business group on board to deliver an overarching strategy. At some companies, the corporate finance team joins in on the fun by setting and levying a

carbon tax that each business group must pay for how much it emits. A corporate sustainability team is often filled with dedicated sustainability roles around data collection and management, disclosure and reporting, stakeholder and program management, policy, and (sometimes) sustainable finance.

- **The Business Group Sustainability Team** must translate the high-level sustainability commitments into something tangible and relevant for its business group. It must work with all of the leaders in the business group to ensure each product and supporting function is clear about (and has data to support) their targets, enablement to build strategies, and program management to ensure the overall work stays on track. Unlike the corporate team, the business group teams have ownership of product portfolios, which gives them opportunities to translate broader sustainability work into more tangible portfolio strategies. A business group sustainability team often has similar types of roles as a corporate sustainability team, with added emphasis on program management and data and sustainability accounting expertise with a background in the product types sold by the business group.

- **The Product Sustainability Team** must then take the work, targets, and context from the business group sustainability teams and figure out how to actually make them *happen* at a product level. If the business group team is the one determining the roadmap for the group, the product teams are the ones that must determine how individual products can be reimagined to achieve the needed sustainability outcome—as well as how it can be done in a way that still delights the customer while meeting financial and timeline targets. They can trust the business group team to help them know how they fit into the bigger picture and then focus their effort on building a product to support it. This tier of sustainability jobs starts to become more filled with people with product leadership over sustainability expertise, but knowledge of sustainability becomes more important as customers and competitors anoint sustainability as a necessary product feature.

- **The Functional and "Execution" Sustainability Teams** are the ones that go out and execute or implement sustainability in support of the product plans. These teams engineer the products, contract with suppliers for the materials to make the products, manage the logistics network to deliver the product, have marketeers to market the product and salespeople to sell it, and more. Doing sustainability work in this tier requires knowledge first and foremost of how to execute the functions, with sustainability often an add-on to the work. So anyone with a skill set in the above could be a candidate to execute sustainability, even if they don't have a formal background in it! That's not to say that dedicated sustainability roles can't flourish at this level. Experts in renewable energy procurement, for example, may live here to decarbonize suppliers and operational facilities.

In other words, it's unlikely that there's a single "sustainability team" at large companies if those companies have made the investment and commitment to become more sustainable, because doing it in earnest requires teams embedded at every layer of the organization. This is probably one of the biggest "aha" moments for people who reach out to me looking for insight on how to get into corporate sustainability.

So, with sustainability teams all over a lot of these large corporations, what does it take for those sustainability teams to be successful? Ultimately, the things that make sustainability successful in a large corporation are (mostly) the same things that make corporations successful in general. Here are a few things that I've found have scaled impact.

Make Sustainability Exciting

Ultimately, at the new scale in which it must now happen, sustainability will be seen as a relatively new work cause for the average rank and file of a company. People who have been in the company for decades are now faced with the notion that they will be asked to support sustainability efforts in some way—and, like any other new initiative they'd be faced with, they need to be inspired to care about it. Thankfully, sustainable causes already have a lot of supporters and can carry their own set of energy. However, it's critical to recognize that the average person (at least today) is more focused on doing well in their own job rather than taking on potential risk by layering a sustainability focus to their work.

To that end, sustainability needs to be presented as an irresistible opportunity that can generate not only top-down buy-in but also bottom-up support. I've seen some sustainability professionals take the opposite approach, where they try to get people to care about sustainability using fear-mongering rhetoric ("If we don't switch to renewable energy at our headquarters, the planet will end!"). Even if this sentiment is true to some extent, within a corporation where sustainability is one of many competing priorities, this is almost never the winning way to get people who are ambivalent to sustainable causes to start prioritizing the work. Instead, these people might shut out the sustainability voices as a depressing (and not career-advancing) distraction from their work that needs to get done.

Taking the approach of "irresistible opportunity," on the other hand, generates positive energy. Using language such as "Sustainability can differentiate our products," "Sustainability can drive revenue opportunities with our customers," or even "Sustainability is a really hard engineering challenge; let's see if we can get creative" can flip sustainability into a career-advancing priority. And since sustainability needs to layer into every part of the organization, building this positive energy will help to break down barriers to collaboration across it.

Make Sustainability Worth It

In addition to bringing inspiration and excitement to the topic, sustainability professionals will ultimately need to operate within the reality of large corporations and their financial targets. Given limited budgets, asking to spend a bunch of extra money on sustainability with no discernible benefit to the company will almost certainly deprioritize the cause in favor of competing interests that would benefit the company. So while the "ROI for the planet" may be crystal clear, it's really critical to frame corporate sustainability work in terms of "ROI for the company" too.

Sometimes you'll be fortunate enough that the ROI is simply "because our CEO said so." But more often than not, your ask to dedicate resources and headcount to a sustainability initiative requires the same financial rigor as anything else. There are a few vectors worth exploring here:

- **Sustainability Can Be Cheaper:** This is typically the most compelling vector as well as the lowest-hanging fruit. In essence, this lever of ROI needs to demonstrate that the higher-polluting alternative is more expensive than a new sustainable alternative. For example, when shipping a product, shipping it via air freight is the most expensive (and highest polluting) option. Unless there is a critical need to ship quickly, shifting your logistics network to favor more ocean or truck freight can both save money and emissions. Another example is by reducing waste. Buying excess materials or goods only to have them go to waste is both environmentally and fiscally unsound. Perhaps there is a way to switch a manufacturing process for a product that creates less scrap material, lowering costs, waste, and emissions in the process. And sometimes, if you're lucky, there's a like-for-like swap for a material or energy source that is just flat-out cheaper. Maybe solar is cheaper on your local grid than coal, for example!

- **Sustainability Can Generate Revenue (and Beat the Competition!):** This is often needed if the cost of a new sustainability initiative is at cost parity or more expensive than a traditional approach. Let's say that you want to use recycled materials in your product or packaging. Unfortunately, many recycled materials are more expensive than virgin materials today (hopefully not for long!). But what if you can show a market study that says that customers would be willing to pay more for a more sustainable product, so increasing costs by 5 percent could increase sales by 10 percent? Or, what if your competitor has just started using this recycled material and is getting a bunch of positive press, so *not* using this material might reduce your market share? Or, even better, what if you have customers *today* who are threatening to go to your competitors because you don't have enough proof of sustainable operations to meet their contractual needs? In a world where more and more purchasing power flows into the hands of customers from generations that indicate more willingness to spend on sustainable brands, making these moves can be a long-term revenue play to build your brand for your future customers.

- **Sustainability Is a License to Operate:** Sometimes sustainability isn't optional at all. Certain geographies, especially Europe, have become incredibly aggressive in regulating corporations to become more sustainable. Being out of compliance with these laws means, at worst, your company would be disallowed to sell into certain markets altogether. A savvy sustainability professional should try to keep an eye on the evolving regulatory regime (and find expert partners that can decode the latest goings-on). Getting ahead of upcoming environmental regulations is a great way to de-risk the business for environmental causes.

- **Sustainability Is an Investor and Employee Mandate:** Ultimately, if you're in a public company, the true "boss" isn't the CEO. It's the board of directors, on behalf of the shareholders—and shareholders are ultimately the ones who determine what the share price of the company stock ends up being. If shareholders are demanding proof of sustainability, not delivering on those demands can lead to stock sell-offs (therefore reducing the price) or, more disruptively, shareholder resolutions. Later in this section, we'll talk about how investors in ExxonMobil took a "provestment" strategy, voting against the company to elect more climate-friendly board members. And on the flip side, just as investors can choose where they want to invest, employees can choose where they want to work. If the employees (either current or potential) start to mandate more sustainable activity, the cost of *not* doing sustainability could be reflected in the cost of losing out on great talent.

Make Sustainability Easy and Clear

Let's assume that we've brought a compelling reason to do sustainability into our company. Now what? How do we pull together sustainability into a cohesive vision and strategy rather than a series of independent activities dispersed across the company, some of which may or may not make a meaningful difference?

At a corporate level, this is where clear and measurable commitments become so powerful. It gives the company a language by which to categorize the potential sustainability opportunities it may want to pursue. At Microsoft, we defined *sustainability* as four things:

1) Becoming Carbon Negative by 2030 (i.e., removing more emissions from the atmosphere than we produce annually)

2) Becoming Water Positive by 2030 (i.e., replenishing more water than we use)

3) Becoming Zero Waste by 2030 (i.e., eliminating waste across operations, products, and packaging)

4) Protecting Ecosystems and Becoming Land Positive by 2025*

You can check out Microsoft's annual sustainability report online to dive deeper into these categories and what they mean in practice.

Zooming in on the company's Carbon Negative commitment, it has a measurable reduction target associated with it: to reduce Scope 3 carbon emissions—the emissions of a company's products and supply chain—by more than half by 2030 from a 2020 baseline. This helped sustainability leads in each business group drive clarity on what needed to be done. First, can we measure our emissions accurately? And second, can we demonstrate that our initiatives reduce emissions in a meaningful way?

This lets businesses create a long-term roadmap that balances all of the investments that are needed to hit the target in a single, coherent view. It's one thing to go tell a product team to "reduce as much embodied carbon as you can." It's another entirely to say, "If you reduce the emissions of your next three products by 20 percent apiece by doing X, Y, and Z, in context with our other efforts, we will be on track to hit our target." This can help the company prioritize investment rather than go on wild goose chases for things that don't matter as much in the end.

Here's why clarity on measurement and roadmap is so important. In late 2023, LEGO ended up walking back its commitment to use recycled plastic in all of its products by 2030.[132] It's not because LEGO decided it didn't care about the environment anymore. On the contrary, the company's newest data suggested that using this particular type of recycled plastic (i.e., old plastic water bottles) would actually *increase* its emissions by 2030. So in this case, something that *sounded* really sustainable (using recycled plastic instead of virgin plastic in its products) ended up having a worse impact on emissions. To LEGO's credit, its bravery in admitting the need to change direction after getting better data is great, and the company has shifted its goal to using different sustainable materials by 2032.

Make Sustainability . . . Boring?

"But wait!" you may exclaiming right now. "Didn't you just tell me to make sustainability *exciting*?" Well, yes, I did! But at the same time, sustainability can't become *too* exciting. Allow me to explain.

One thing that I've found sometimes is that, within a company, a bunch of energy is generated to do something on sustainability. An executive has bought into a really cool idea, the employees are excited about implementing it, and then they go off to try to execute it. However, they try to go and execute it outside of the existing processes that the business uses to operate, ultimately slowing down and draining the initial enthusiasm that came with the idea.

Many times, sustainable initiatives within corporations fail because they are treated as an "exciting other," an idea born outside of the system rather than an idea run through the system. A lot of energy is then wasted to get people excited about an idea that nobody knows how to actually implement or do.

Instead, sometimes it's worth making sustainability "boring"—in other words, embedding it into the existing business systems in the same way as any other type of initiative or investment. That can mean having sustainability show up as a line item in a spreadsheet, a single slide in a massive product deck, or a five-minute agenda item in a two-hour operations review. Not to say that

---

[132] The LEGO Group, "The LEGO Group Remains Committed to Make LEGO® Bricks from Sustainable Materials," September 27, 2023, https://www.lego.com/en-us/aboutus/news/2023/september/the-lego-group-remains-committed-to-make-lego-bricks-from-sustainable-materials.

a two-hour operations review can't be exciting, but generally the more that sustainability can be "just another thing" for a business to manage, the more likely it is to succeed by using the same infrastructure that everything else in a business needs to do to get done.

Ultimately, across all of the above, my hope is to encourage you that you're not that far away from making sustainability part of your job—and that your knowledge of how to do your existing job is a critical asset to make it happen. You just need to make the opportunity yours.

Discussion Points

1) How could parts of the above framework help you in how you do your day job, even outside of sustainability?

2) If you are interested in corporate sustainability, what layer appeals most to you? How does this intersect with your North Star?

3) Where have you seen sustainability initiatives in your workplace that didn't work? Based on the advice in this section, what could you do to make those initiatives successful?

Chapter 11: Your Investments and Savings

The way we make money to survive and thrive isn't just limited to the physical labor we put in every day. Some of us are also fortunate enough to invest some of our money. The old adage is to "make your money work for you." But why can't we amend that philosophy to consider how our money can work for the planet as well?

I want to differentiate between sustainable investment and donations, which I will cover as part of a later chapter focused on community and volunteering. For this section, we'll take the more classically capitalist approach of investing money to achieve ROI. But investment to this end is starting to align more and more with supporting sustainable causes and beautifully so.

Investing Sustainably

One type of sustainable investing that's been in the news recently is ESG fund investing, especially in retirement plans or a 401(k). In short, it's an investment philosophy that qualifies which stocks to purchase for a fund (e.g., an index fund, an exchange-traded fund or ETF) based on a set of criteria that includes environmental, social, and governance responsibility in addition to financial health. Many ESG funds will usually start with the basics—excluding stocks likely to have low ESG scores such as from gun or oil/gas companies—and then expand to weigh their portfolios toward companies with higher ESG scores. While a very imperfect vehicle today (you'd be surprised at how many "bad" companies still somehow sneak into ESG funds), it's a way to send a market signal that you're willing to direct your investments away from bad actors and toward companies that show commitment (at least outwardly) to sustainability.

Thankfully, investing in ESG funds has often not come at personal expense and can sometimes even yield a better return for investors and outperform the broader market. Whether this performance is directly tied to a company's willingness to focus on ESG is largely up for debate. It is probably more likely tied to the fact that most ESG funds tend to be overweight on tech (or tech-adjacent) stocks, which have also generally outperformed the market. As an example, FITLX, a Fidelity-offered ESG fund for US stocks, has Microsoft, Nvidia, Google, and Tesla as its top four holdings, representing over a quarter of its total holdings.

Companies (and their boards of directors) are paying very close attention to this ESG investment phenomenon, and many investors in earnings calls are pushing companies heavily to increase their ESG performance so that they can attract more investment from retail (i.e., average people) and institutional (i.e., banks, pensions) investors. In this way—and in addition to potentially making you MORE money than you would otherwise—you are helping reward companies that are taking a more active leadership role in sustainability. You're also driving financial incentive for laggards to follow suit. Though, to play devil's advocate, we can't yet conclusively say whether investing in ESG funds has led to quantifiable reductions in emissions.

Beyond ESG funds, there is the emergence of a new type of investing I like to call "provestment," which looks to wield shareholder power to influence positive change—even at dirty companies. For example, new types of sustainability ETFs—like the VOTE ETF devised by asset management firm TCW—will actively vote for social and environmental causes in company board resolutions on behalf of your shares. Provestment investor pressure, when applied collectively, can drive real behavior. In 2021, ExxonMobil, one of the world's largest polluters, suddenly found itself losing votes on who to nominate to its board of directors. Instead, sustainability-minded activist investors rallied enough support to *outvote* Exxon-Mobil and instead vote in three sustainability-interested board members.[133] The power of the dollar, when collectively applied, can be quite powerful indeed.

> Investing in ESG funds or using provestment to vote with your shares can pressure companies into increasing their focus on sustainability.

Ironically, this newer concept of provestment to become a shareholder of a "dirty" company to drive change is, in some regards, directly opposed to the philosophy of traditional ESG investing. You have likely heard of the opposite concept of "divestment," which became in vogue especially

---

133 Matt Phillips, "Exxon's Board Defeat Signals the Rise of Social-Good Activists," *New York Times* (online), June 9, 2021, https://www.nytimes.com/2021/06/09/business/exxon-mobil-engine-no1-activist.html.

at college campuses in the 2010s. This concept, in short, is to pressure institutions (and individuals) to divest from their existing holdings of fossil fuel and other dirty companies so that their investments don't support and won't profit from those companies. In theory, divestment applies similar market pressure on share price as ESG investment does—but on the side of "taking money away from the bad guys" rather than "giving money to the good guys."

Just as ESG investing has not yet shown to conclusively reduce emissions, divestment has not been found (at least not yet) to materially impact the behavior of the companies being divested from, which are typically the ones most diametrically opposed to climate action.[134,135] Quoting Michael O'Leary, managing director of the activist investor firm that led the charge to unseat ExxonMobil's board members, "For one investor to sell a share, another must buy it. Period. An unhappy shopper can complain to management or take his business elsewhere, but an unhappy shareholder simply replaces herself with someone who cares less."[136] Ironically, divesting risks removing the voice of ESG-minded shareholders who may care about trying to vote to hold these companies more accountable. In this way, divestment could potentially make dirty companies even *more* likely to ignore ESG! From an environmental point of view, provestment means you can still make an impact while holding onto dirty stocks—but only if you exercise your voice as a shareholder to demand positive action. If you don't (or can't), or philosophically want autonomy to fully align *all* your (or your institution's) investments with your values, then consider divestment.

Saving Sustainably

In addition to where you place your money for active investment, there's the consideration of where you keep your passive money (e.g., banking, credit unions, bonds) and your spending money. Not all banks are created equal when it comes to what they do with your money. In some cases, even if you are not investing in fossil fuel production, your bank (especially if it's a large multinational one) might be lending your money to fossil fuel companies to fund their operations. Credit unions, which are typically more focused on local, mission-based lending, can give you a better picture of what your money goes to if this is a concern.

---

[134] Adam Aston, "Why Divestment Doesn't Work—and Just Won't Die," GreenBiz, December 15, 2021, https://www.greenbiz.com/article/why-divestment-doesnt-work-and-just-wont-die.

[135] Alexander Gelfand, "Why Divestment Doesn't Hurt 'Dirty' Companies," Insights by Stanford Business (blog), Stanford Graduate School of Business, October 27, 2021, https://www.gsb.stanford.edu/insights/why-divestment-doesnt-hurt-dirty-companies.

[136] Aston, "Why Divestment Doesn't Work."

Thankfully, there is a set of emerging banks that are sustainability-forward as well as a set of existing banks that are willing to offer more attractive eco-friendly options. The best way to tell if your money will go to a good cause is to investigate the types of personal and business loans that your bank or credit union tends to offer. Some may specialize in home solar loans, for example, which gives a good indication that your money will be used to actively help others purchase solar panels. Typically, smaller, niche banks and credit unions will afford you much more transparency on what they spend your money on—and being cognizant of their investment strategy can help you ensure your money is being used for good. One such niche bank, Atmos Financial, estimates that for every dollar saved, nearly two pounds of emissions are avoided annually by virtue of lending out bank members' deposits to finance solar loans.[137]

> Saving your money at sustainability-forward banks or credit unions can fund sustainability activity like solar loans.

Spending Sustainably

The final piece of this puzzle is the credit or debit card that you use to take money out of your bank. If you can believe it (which, by now, I imagine you can), even this choice ends up making a potential impact. Many cards offer perks to you for using them, but those perks don't always add up to a positive impact. For example, one of my earliest credit cards would accrue me points that the card issuer would mostly encourage me to use for going on cruises, which are a problematic type of vacation to support as mentioned earlier.

> Using a green credit or debit card can tie your spending to impact.

Thankfully, a new breed of credit and debit cards are emerging that tie your spending to im-

---

[137] "How Atmos Calculates Carbon Impact," Atmos Financial (blog), last modified June 14, 2024, https://www.joinatmos.com/blog/how-we-calculate-carbon-impact.

pact. Many of these will give greater cash back for "sustainability" themed purchases, such as bus passes or e-bikes. Some will plant trees every time you spend enough money or donate to other charitable causes. Some cards are made from recycled plastic or even wood. Others will even aim to give you a perspective on your personal emissions based on your spending habits! A quick internet search for "sustainable debit and credit cards" will yield a number of really exciting emerging options that are worth considering.

Cryptocurrency

I would be remiss if I didn't address one of the newest breeds of investing, namely crypto. If there is one piece of advice I would leave you with on sustainable investment, it is this: Avoid Bitcoin in favor of other crypto.

The way the Bitcoin system works, put very simply, is that individuals (aka "miners") will race to solve complex problems as fast as possible, most often by using as much electricity as possible with computers specifically built and optimized to do so. The Bitcoin blockchain—the ledger of decentralized transactions that makes trading Bitcoin possible—rewards this by awarding Bitcoin to the miners, who help maintain the blockchain network by virtue of this work. This system is called a proof-of-work system; whichever miner's equipment worked the hardest and the fastest gets the reward.

Though data varies, generally the result is that the amount of energy used to manage the Bitcoin network today (at the time of this writing) is *larger than the entire country of Argentina's annual electricity usage*. And while data is scarce on the impact of water to produce this electricity and cool the machinery producing Bitcoin, a doctoral candidate at Vrije Universiteit in Amsterdam estimated that in 2023 Bitcoin used *more water than the entirety of New York City*.[138]

On a per-transaction basis, as estimated by the Bitcoin Energy Consumption Index in November 2023, a single Bitcoin trade can take almost the same amount of energy as your entire house does in a month.[139] Even the computer I am writing this book on took less energy to make and use than had I made one Bitcoin trade. And that is not to mention the embodied emissions and waste associated with the construction of mining facilities and manufacture of mining hardware, which are often overlooked and undercounted in measuring Bitcoin's total environmental footprint.

It's hard to estimate specifically, but per the same source, the energy required to trade one Bitcoin (not to mention even mining one) is equivalent to

---

138 Alex de Vries, "Bitcoin's Growing Water Footprint," *Cell Reports Sustainability* 1, no. 1 (2024), https://doi.org/10.1016/j.crsus.2023.100004.

139 "Bitcoin Energy Consumption Index," Digiconomist, accessed July 2024.

almost *one million* credit card transactions. Similarly, the amount of water used to trade *one* Bitcoin is roughly equal to a backyard swimming pool. In this case, looking at the bottom-up, per-transaction impact of Bitcoin helps underscore how inefficient proof-of-work systems are as compared to traditional forms of value exchange; Bitcoin's underlying protocol fails the "Reduce" part of "Reduce, Reuse, Recycle."

Even if some Bitcoin miners aim to buy up excess renewable energy capacity to offset the energy portion (as well as reduce some of the water usage in generating the electricity), though that may reduce near-term impact and help stabilize the grid, ultimately it takes that excess clean energy off the market for future buyers, pushing others to use less renewable energy in the interim.*

The exception to this would be if a miner signed a clean energy PPA that achieves additionality, i.e., guarantees the net new creation of renewable energy production. These types of deals can help accelerate the clean-energy transition.

Thankfully, if you still want to be involved in crypto trading, there are alternative blockchain variants that are less environmentally taxing. Ethereum, the second largest cryptocurrency, shifted to what's known as a proof-of-stake blockchain model, which does away with the traditional mining component of cryptocurrency and instead validates trades in an over 99 percent(!) more energy-efficient manner.[140] (I will avoid going into too much depth about how this works for your sake, reader.) That being said, while proof of stake is an *enormous* improvement over proof of work and Bitcoin, it is still more energy-intensive than traditional trading and card apparatuses.[141] But for those who wish to remain in crypto and Web3, just make sure you confirm that the crypto that you are trading is managed and run on a proof-of-stake blockchain—and run away if you learn that it is a proof-of-work blockchain.

> Avoid Bitcoin and "proof of work" blockchains in favor of other cryptocurrencies built on "proof of stake" blockchains.

---

140 "Ethereum's Energy Expenditure," Ethereum.org, last modified October 24, 2023, https://ethereum.org/en/energy-consumption/.

141 "Ethereum Energy Consumption Index," Digiconomist, accessed July 2024, https://digiconomist.net/ethereum-energy-consumption.

Discussion Points

1) Have you ever considered how your investments and savings could have a sustainability impact? Are there examples that surprised you?

2) Do you know what your bank uses your money for? What would your "ideal" bank have as its values for investing?

3) Does looking at impact on a per-transaction basis provide more meaningful insights than looking at it on an overall basis? Which do you find more meaningful? Why?

Closing Part Two

To close this chapter on our "professional" lives (i.e., our work, careers, and investments), I hope it has become clear that **every job can be a sustainability job**, and every financial interaction can be a sustainable interaction.

Whenever I speak with individuals looking for career advice on how to get into sustainability, I often sense frustration from their prior experiences. Perhaps they have applied to hundreds of sustainability jobs and gotten no hits. Maybe they feel they don't have relevant skills and are weighing the difficult decision to go back to school to get an expensive degree that they feel could help them get into the space. Many times, they have no idea even where to get started—or what a sustainability job could even look like.

To quote Drew Wilkinson from earlier in this section, bringing sustainability into your career is often not a light switch but rather a dial. I feel that a lot of this angst comes from the assumption that getting into sustainability needs to be like a light switch and happen immediately. In reality, most careers are decades long, and turning that dial to increase the amount of sustainability in your career can take years. I needed to turn my own dial for seven years in my career before landing my first job with "sustainability" in the title.

So if it doesn't happen right away, don't get discouraged. Keep your patience, and keep turning that dial. If all of us turn our dials even a little bit, the collective impact could become incalculable. Remember, even if it seems like it sometimes, the climate crisis is not all-or-nothing. There is a large difference between warming by 1.5 degrees Celsius and warming by 4 degrees (if we all give up). Every little bit matters!

Here's one final point that I'd like to make, which we'll expand on in the next section: There is also a class of non-sustainability jobs and activities that have a surprising correlation to generating sustainable outcomes, some of which you might already be a part of. For example, working as a mental-health counselor can lead to a positive impact on the planet, not just on people. Further, the economy needs people in traditional "non-sustainability" jobs to keep functioning, which is critical to enable sustainability work to occur in the first place. As you read the next part on what we can do in our "political" lives, consider carrying the lens of your career with you to understand the power of sustainability-adjacent work for saving the planet.

Discussion Points

1) What are the top things you could do in your career to increase your impact? What might prevent other people from doing the same? Is there a way to make it easier for them to do so? How do we bring others along with us?

2) Why do you think people have historically underappreciated the role of traditional careers in helping drive sustainability? What could help people understand the potential impact of the place where they are spending a third of their time?

3) Who should you include as mentors and connections as you build a network of professionals to help "turn the dial"? What skill sets and roles do these people have that you could learn from?

PART THREE: OUR POLITICAL LIVES

Chapter 12: Our Impact Is Directly Proportional to Our Sphere of Influence

Now that we've covered some of the most impactful things we can do in our personal and professional lives, the last—and most impactful—category focuses on what we can do in our political lives. Ultimately, **our impact is directly proportional to how wide our sphere of influence is**. Generally, our communities are the largest group of individuals that we have levers to make a difference for and with. Therefore, **the impact of how we vote to guide these communities can exceed the impact that we could make individually in even a full year—or more!**

In our personal lives, we are generally making decisions just for ourselves and our families. The magnitude of our personal sphere of influence is in the order of ones to tens of individuals. While we have ultimate control over what decisions we make individually every day, that unilateral control tends to be very limited in scope.

In our professional lives, our sphere of influence grows to our customers and coworkers. And except for the rare roles and companies that are truly massive in scale, most of our jobs give us a magnitude of influence in the order of hundreds to thousands for our coworkers and thousands to millions for our customers. We likely have less control over the ultimate direction of our work, but small impact can multiply over many more coworkers and customers.

But for most of us, the single biggest thing that we are connected to—and have some element of influence over—is our communities. These range from your immediate family and friends to your town or city and to the state and

> Getting involved in your community is often the largest opportunity for impact you can have.

country you live in. The opportunities we have for impact can range from the hyper-local (e.g., volunteering at a nearby park or at mayoral elections) to the national or international stage (e.g., through the nonprofits or federal policies we support). Compared to the "impact sphere" of our personal lives (ones to tens of individuals) or our professional lives (hundreds to thousands of individuals), our impact in our political lives can scale to the size of our nation itself (millions to hundreds of millions of individuals).

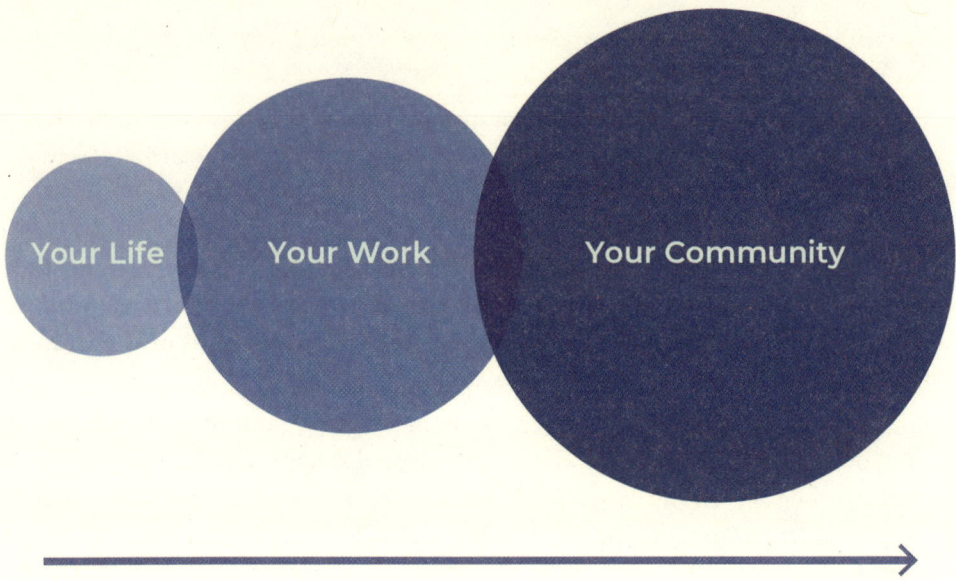

At a high level, as we draw parallels between our "personal" emissions pie chart and our "professional" emissions pie chart, we can also look at the overall emissions profile of our country to get a sense of where our biggest decarbonization opportunities lie in our broader communities. We've already seen this chart earlier in the book, but it bears restating here. There is a three-way tie between emissions from industry, from transportation, and from residential and commercial buildings, with agriculture rounding out the remaining 10 percent of annual US emissions.

WHAT WE CAN DO

TOTAL U.S. GREENHOUSE GAS EMISSIONS BY ECONOMIC SECTOR AND ELECTRICITY END-USE

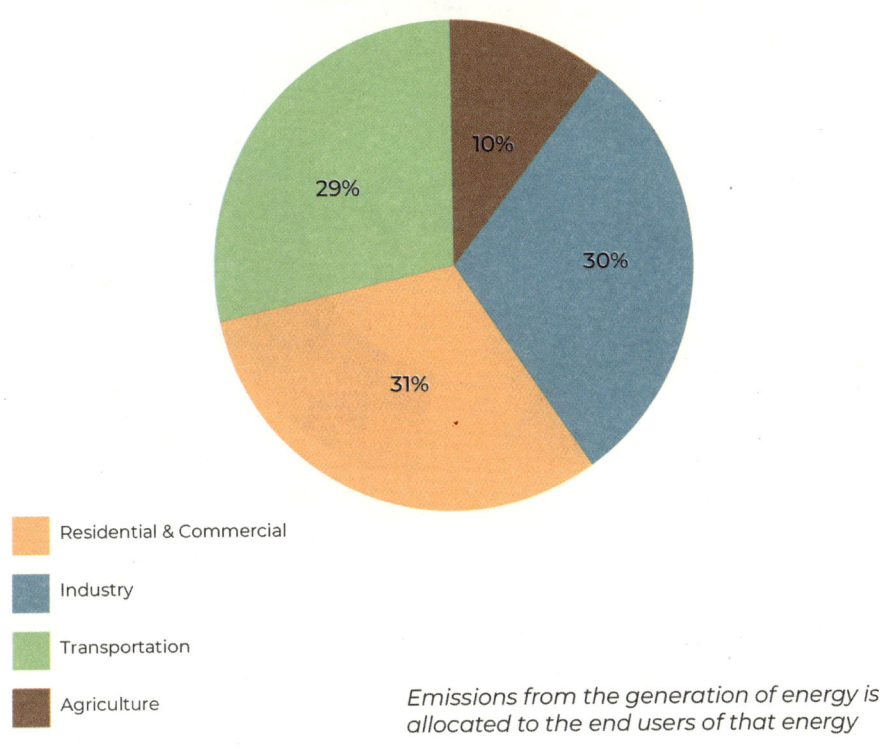

Emissions from the generation of energy is allocated to the end users of that energy

Source: United States Environmental Protection Agency, *Inventory of U.S. Greenhouse Gas Emissions and Sinks*, https://www.epa.gov/ghgemissions/inventory-us-greenhouse-gas-emissions-and-sinks.

Drilling down even further, some cities also publish their own sustainability reports and emissions pie charts. For example, the city of Seattle (where I wrote most of this book) has an interactive dashboard that shows the breakdown of emissions for activities within the city limits from 2020. Each region will reflect a different set of activities and, therefore, a different emissions profile. Because Seattle is a large aviation hub and much of its electricity is generated from hydroelectric power, most of its emissions come from transportation and industry rather than from buildings.

EMISSIONS FROM THE CITY OF SEATTLE

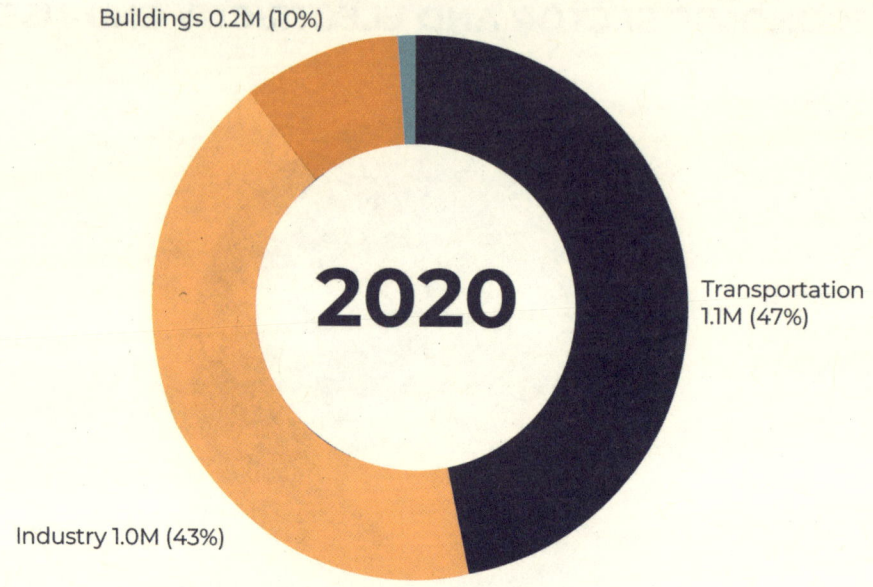

Source: "Understanding Our Emissions," City of Seattle Office of Sustainability and Environment, https://www.seattle.gov/environment/climate-change/climate-planning-and-data/performance-monitoring.

There are many different ways to tackle such a multifaceted opportunity. In many ways, these actions in front of us can also be some of the most deeply personal ones. It asks us to reflect on what is more important to us from policy, impact, and focus perspectives. For instance, "Is it more important for me to adopt a local highway and clean up litter on my commute?" Or, "Is it more meaningful to donate to animal conservation nonprofits, even if the animals being conserved are far away?" Or, "What kind of policies do I want to prioritize with my votes when I'm stuck between only two viable candidates for election, each of whom may support some but not all of what I value?"

The good answer is that most (if not all) action is good action, and you can't go wrong by getting more engaged in your community in a way that's conducive to amplifying impact and bringing others on the journey. What we will focus on now will be an attempt to bring clarity and structure to the types of community impact available to us so that we can make meaningful decisions about what to support based on our personal values.

Discussion Points

1) Think about which communities you belong to today. How could each one contribute to a more sustainable future?

2) What kind of positive impact is most fulfilling to you? What is most relevant to your community?

3) Does your community have its own sustainability report? If it does, read it and reflect on what you think could be some major ways to help reduce your community's emissions. If not, what do you suspect could be your community's biggest opportunities for change?

Chapter 13: Volunteering and Nonprofits

For those of us who are fortunate enough to be in a position to donate our time or money, it can become daunting to understand which nonprofits and volunteering opportunities are most impactful. For every impact area—say, water conservation—you can find dozens upon dozens of nonprofits that claim to offer the most impact for your dollar or time. Just like private companies, many nonprofits will invest in marketing and newsletters to draw more income—and less scrupulous ones may overstate their impact.

While there is no perfect barometer for determining what to support, thankfully there ARE some resources available to help break this down. Personally, I tend to enjoy Charity Navigator (a nonprofit itself) as an independent assessor of how much of a nonprofit's income goes directly to supporting its mission. However, more hyper-local opportunities or newer nonprofits might not have the size and scope to be independently assessed.

My general rule of thumb is to move beyond "what does this nonprofit *claim to do*" to "what has this nonprofit *done*." Good nonprofits tend to publish impact reports, where you can assess whether what they've worked on looks and sounds impactful (assuming they are being genuine and not over-marketing their impact). In some cases, maybe you've seen some of their impact in your community firsthand, which is a great barometer.

That's not to say that nonprofits that don't publicize their impact can't be meaningful. Maybe they are new or small and are scaling toward a point to increase their impact. In those cases, it's good to get to know the nonprofit directly. Maybe one of your friends is directly involved and can speak to what they're working on. But generally, hyper-local nonprofits probably have less global impact and more local, pointed impact, ideally filling an unmet need. These nonprofits are often the ones where your volunteering and donations make the biggest relative impact, and your individual decision to donate could be the difference between the nonprofit persisting or closing up shop.

As for what kind of nonprofit or volunteering opportunities to support, it ultimately comes down to what is most personally fulfilling and enriching to you. Choosing to support plastic waste prevention over tree planting, for example, is a choice between two good things. It just depends on whether plastic waste or climate change is your bigger personal priority. And if you want to maximize impact within your preferred focus, I will attempt to give a view as to which areas—some obvious, others surprising—can best advance your cause.

As a rule of thumb, local causes will have a more immediate and visible impact, but the total impact is usually smaller in aggregate. State- or nationwide causes will have a less visible near-term impact locally but the potential for a larger overall impact in the medium term. And while global causes are the least likely to result in visible local change, they offer the potential for the largest scale of impact over the long term.

Near-Term or Local Impact

If you are most interested in getting involved with local nonprofits, having an immediate, tangible impact is likely a key priority rather than large-scale environmental transformation. So for near-term, local impact, ultimately the decision of what you want to get involved with should be a reflection of what you think would yield you the greatest satisfaction and accomplishment within your community. Perhaps there are volunteer groups that are focused on tree planting or litter cleanup in a park. Perhaps there are opportunities to provide outdoor activities to your local schools, worship communities, or social groups to engender a greater love and appreciation for the environment. Or, perhaps there are homeless services you can provide to house local individuals (which we'll talk about later in regard to its opportunity to benefit the environment).

I'd like to draw attention to one type of local giving that is often overlooked when it comes to environmental impact (and generally): giving to Native American nonprofit causes. As the original stewards of the environment in North America, many nonprofits focused on Native American issues and rights are often inexorably tied to environmental reclamation and ecological preservation. I'd encourage you to get informed on the work of your local tribes and tribal nations.

Ultimately (and pragmatically), the environmental causes that you support locally should be a reflection of your value system and your desire to better your community rather than an aim to maximize impact. If I would caution against anything, it would be to consider ways to volunteer or donate locally in a way that doesn't have adverse effects, such as whether you can have alternatives to boxed lunches with plastic chip bags and plastic water bottles for volunteers.

WHAT WE CAN DO

Medium-Term or National Impact

There are many state and national causes that are high impact to give to. These range from "urban reforestation" nonprofits and American wildlife conservation to natural land, forest, and park preservation. Some of these nonprofits often work at global levels, too, aiming to engage with emerging countries on ecosystem preservation. These are all generally great organizations and missions to consider supporting.

But to expand our focus beyond "obvious" nonprofits that are wholly dedicated to sustainability and conservation, I'd like to share an example of a surprising but highly impactful cause to support for the environment: **supporting better access to voting and the expansion of voting rights**.

Anything that makes voting easier, more enticing, and more representative are opportunities to green the planet. These include voter registration nonprofits, get-out-the-vote nonprofits, ranked-choice voting nonprofits, and voting rights expansion nonprofits. While some might go even further and support individual candidates' campaigns, even just the agnostic support of voting access and voter engagement—no matter who it targets—has the opportunity to benefit the environment significantly, the means by which we'll get into later in this chapter.

> Supporting voting rights and access can be one of the most impactful 'sustainability-adjacent' ways to support sustainability.

In America, this is generally due to a few reasons. First, it ensures that more Americans have their voices heard and their votes counted on a generally *popular* policy position. For a topic like climate change and sustainability, if we look again at the Yale study on "Global Warming's Six Americas," only 22 percent of respondents are "doubtful" or "dismissive" of climate change.[142] (Refer to the Introduction for more information.) Another poll from the Pew Research Center in 2022 found that 69 percent of Americans supported steps for America to become carbon-neutral by 2050, against only 28 percent who opposed it.[143]

---

[142] "Global Warming's Six Americas," Yale Program on Climate Change Communication.

[143] Alec Tyson, Cary Funk, and Brian Kennedy, "Americans Largely Favor U.S. Taking Steps to Become Carbon Neutral by 2050," Pew Research Center, March 1, 2022, https://www.pewresearch.org/science/2022/03/01/americans-largely-favor-u-s-taking-steps-to-become-carbon-neutral-by-2050/.

Climate-friendly policies are fast becoming universally supported, even across party lines. However, that does not necessarily translate into it becoming a priority for both parties . . . yet.

Second, at least at the time of this writing, more disenfranchised voting groups (often minority groups) or disengaged voting groups (often younger voters) tend to vote disproportionately for candidates who support climate-friendly policies. It makes sense; younger voters have the most to lose from climate change (as older voters likely won't be alive by the time the worst effects of it come to pass), and minority groups are often the demographics that are most disproportionately impacted by climate change, as they often live in less climate-resilient parts of cities for a long list of unfortunate historical reasons, such as redlining. Think back to Hurricane Katrina from 2005, which decimated the Black-majority Lower Ninth Ward in New Orleans. Environmental justice looks not just to save the planet but to also acknowledge and address the disproportionate impact that climate change can have on marginalized communities.*

Environmental justice, by the way, is a great category to consider supporting or volunteering for as well. I highly recommend reading more about the movement from other authors who can do it better justice than I can (no pun intended).

Long-Term or Global Impact

If you wish to prioritize investment in long-term or global sustainability solutions, recall some of the lessons discussed earlier in this book. A significant chunk of the emissions- and waste-producing activity around the world comes from making things, not just using things. Keeping this framework in mind can help to significantly expand the type of causes we support.

In particular, the benefit of global nonprofits and NGOs is the capability to (at least theoretically) put pressure on supply chains that span the globe. While America unfortunately has one of the highest emissions-per-capita rates in the world, it also outsources a significant amount of its supply chain globally—especially to China, the country responsible for the largest amount of greenhouse gas emissions annually (over twice those of America, per analysis from the European Union).[144] NGOs and nonprofits that focus on conservation in these countries—or advocacy for cleaner forms of energy, manufacturing, and products there—are intriguing ways to think globally with your giving. (As long as the NGOs don't engender negative unintended consequences, of course. Always do your research first!) This also extends to things like protecting the

---

144 Joint Research Centre, *CO2 Emissions of All World Countries—JRC/IEA/PBL 2022 Report*, EUR 31182 EN, Publications Office of the European Union, Luxembourg, 2022, https://doi.org/10.2760/730164.

Amazon rainforest and providing protections against the expansion of cattle farming and exports (the "beef" supply chain).

Another intriguing lever we have globally to combat climate change is to support women's education—or even education generally—in developing countries. While this may seem like a significant logical leap to some people, there's some surprising linkage between women's education and reduced emissions. Per the World Bank and many other sources, the higher the level of education for women, the fewer the number of children they tend to have (sometimes by half, other times by more).[145] Generally, no matter where you are in the globe, a more educated and empowered female population tends to correlate with a reduction in births per person and an increase in the average age of childbirth (including in America, per data from the US Census).[146] As we saw in part 1, smaller and "older" families tend to be one of the most significant ways to reduce emissions in the long run by combating overpopulation.

It's an investment that will take decades to show meaningful effect on the planet. But the benefit to the lives of women globally in the near term shouldn't be overlooked, either!

It's also worth asking, "What about donating to receive carbon offsets?" Carbon offsets started to become in vogue over the last decade as a way to pay or donate for someone else to "do the good work" to offset your emissions. In addition to stand-alone companies that offer this, some companies (like certain airlines) will ask you for a couple of extra dollars to offset your emissions. In many ways, these are attractive because you don't have to alter your behavior if you're willing to pay extra to offset your emissions.

However, carbon offsets are not the silver bullet that we had once hoped they'd be. For starters, they are not always tied to directly removing the carbon emitted. Instead, they are often predicated on the notion of preventing future emissions. Buying one metric ton's worth of emissions offsets doesn't guarantee that you'll undo the ton you'll try to offset. It just prevents another ton from entering the atmosphere, like through protecting forests from being cut down.

At the most cynical, there is the unfortunate reality that there are a number of bad actors in the carbon offsets space due to loose standards and enforcement for offset providers to guarantee projects actually happen and their impact lasts. In theory, I could buy a parcel of forested land, claim it is at risk of being cut down, accept payments to prevent any trees from being felled, and then certify those payments as "carbon offsets." Some companies, such as Del-

---

145 Elina Pradhan, "Female Education and Childbearing: A Closer Look at the Data," World Bank Blogs, November 24, 2015, https://blogs.worldbank.org/en/health/female-education-and-childbearing-closer-look-data.

146 Gretchen Livingston, "For Most Highly Educated Women, Motherhood Doesn't Start until the 30s," Pew Research Center, January 15, 2015, https://www.pewresearch.org/short-reads/2015/01/15/for-most-highly-educated-women-motherhood-doesnt-start-until-the-30s/.

ta Airlines, have actually retreated from offering customers the opportunity to offset their emissions as a result of a class-action lawsuit suggesting that these offsets were misleading to customers.[147]

That's not to say that carbon offsets can't have a lot of value. On the contrary, carbon offsets (and their more expensive and rigorous cousins, carbon removals) can be a critical way to protect, preserve, and expand our environmental safety net that already exists. But if you want to purchase credits to offset your activities, make sure you do a lot of research to make sure the offering organization—and the specific offsetting projects it is working on—is legitimate.

Discussion Points

1) Which nonprofits matter most to you? Why?

2) What other "sustainability-adjacent" missions could you support that could generate a positive impact on the planet?

3) What role should carbon offsets and carbon removals play in our overall climate strategies?

---

147 Zachary Barlow, "Delta Lawsuit Argues That Carbon Offsetting Won't Fly," PracticalESG.com, June 2, 2023, https://practicalesg.com/2023/06/delta-lawsuit-argues-that-carbon-offsetting-wont-fly/.

Chapter 14: Sustainability Policies to Support

Every Vote Can Be a Sustainability Vote

When it comes time to exercise your voting rights, you are wielding perhaps the most powerful tool available to you to support and champion sustainable causes. This can be at the national, state, city, or local level—down to even your town's or city's school board elections! But the most important takeaway for the climate is this: Up and down the ballot, elections matter, and your voice in those elections can make more difference than almost any other thing you do. **And just like how every job can be a sustainability job, every vote can be a sustainability vote.**

A great example of this power can be seen in the passing of the Inflation Reduction Act of 2022, which was widely hailed as the largest climate-friendly investment bill the United States has ever passed. It put aside $369 billion for investment in things like clean energy and EV infrastructure, as well as tax rebates for individuals to lower the emissions of their homes through electric heat pumps, energy-efficient appliances, and rooftop solar panels. The bill was expected to drive US greenhouse gas emissions in 2030 to levels 40 percent(!) lower than they were in 2005, when they were at 7.5 billion metric tons.[148]

This bill originally passed with razor-thin margins and would not have been possible if climate-supporting politicians were not in positions of power to make it happen and reflect the will of their constituents. All else being equal, the originally estimated decrease from 7.5 billion metric tons to 4.5 billion gives us around 3 billion metric tons of emissions that could be avoided in 2030—roughly the same as growing 50 billion trees for ten years—thanks to the power of voting. If the bill is amended or rescinded before 2030, this impact will be lower but still significant thanks to the money already invested in the few years since its passage.

Very crudely, if you divide the originally estimated reduction by 81 million,

---

148 United States Environmental Protection Agency, *Inventory of U.S. Greenhouse Gas Emissions and Sinks*.

which is the number of people who voted for former president Joe Biden (who signed the bill into law), you get a maximum per-vote impact of around 37 metric tons of emissions. In other words, **the positive sustainability impact of a vote in 2020 had the potential to be roughly the same as two years' worth of an average American's emissions**.

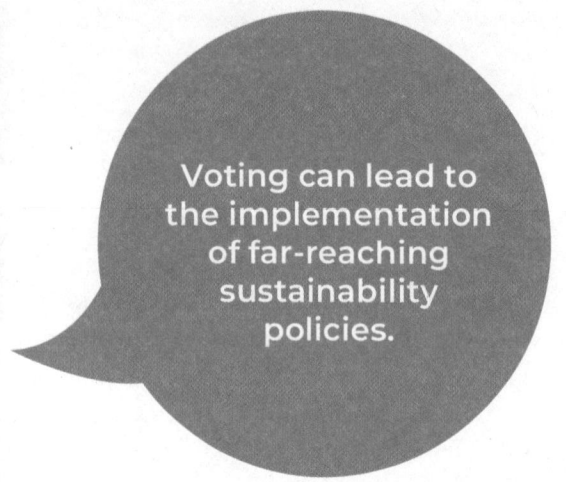

Voting can lead to the implementation of far-reaching sustainability policies.

However, the challenge with "quantifying" the power of a vote like we've done above is that, unlike more tangible things you can do in your life or at work, the impact of a vote is not realized if the vote fails or may be reduced if future voters choose to upend the original vote. And sadly, in today's world, protecting the environment has gone from being somewhat bipartisan and noncontroversial, where environmental bills could get support from both parties, to being deeply polarizing at a national level. People often forget that two of the biggest environmental policy wins in the last half century—the creation of the EPA and the signing of the Montreal Protocol, which is closing the hole in the ozone layer—were both led and championed by Republican presidents (Nixon and Reagan, respectively).

However, the reality of the modern national political landscape of America, at the time of writing, is one of two opposing worldviews: one that believes climate change is real, manmade, and requires action to combat it and one that generally doesn't.

So is the answer just to "vote blue no matter who" if you want to support climate-friendly policy? Well, not exactly. Climate policy and political dynamics are not monolithic and can change over time. Polarization on climate at a national level does not necessarily translate into state or local politics, especially if the environmental policy is more localized as well. A great example of this can be seen in the state of Minnesota, which in 2024 saw almost 80 percent of voters approve an amendment to its state constitution to fund the protection of water, conserve wildlife, improve air quality, and expand access to parks and trails.[149] Localized environmental policy can often achieve broad bipartisan support and be championed by candidates from either side of the aisle.

---

149 Tony Kennedy, "Minnesota Voters Overwhelmingly Approve Renewal of Dedicated Lottery Funding for Environment," *Minnesota Star Tribune*, November 6, 2024, https://www.startribune.com/voters-to-decide-fate-of-minnesotas-dedicated-lottery-money-to-environment/601174846.

While it is not inaccurate to say that, in general, left-wing candidates tend to run on more climate-friendly platforms than right-wing candidates, it is still worth understanding which policies individuals on both sides of the aisle look to support before casting your ballot, especially in regional and primary elections. An ever-growing number of politicians on both sides of the aisle recognize climate change and want to do something about it, and as we move further into the future, it is entirely possible that we'll return to an era when fighting climate change becomes bipartisan again.

In some cases, certain right-wing policy points can be surprisingly climate-friendly, even if they are not explicitly billed as such, which I'll detail shortly. To echo that (and to put my own voting record on display), I have voted for both Democrats and Republicans for climate reasons. And while policy is *really complex* (and I am not claiming to be an expert on policy), I will at least detail some common policy positions that, at least on the surface, look to help drive more sustainable outcomes.

Discussion Points

1) Why do you think sustainability has stopped being a bipartisan issue at the national level? What do you think could help solve this?

2) Think back to some of your most recent regional or local elections. Has sustainability been a part of any candidate's platform? What have they focused on in particular?

3) What role, if any, do other political parties in America besides Democrats and Republicans (e.g., the Green Party) play when it comes to advancing climate policy?

The "Obvious" Policies

Let's go back to two of our core tenets so far:

1) Making things is often far more harmful than using things.

2) Our best way to mitigate this is to reduce, then reuse, then recycle.

Let's also think back to *where* emissions come from in America: industry (30 percent), transportation (29 percent), commercial and residential (16 and 15 percent apiece), and agriculture (10 percent). If we think about sustainability policy in this context, there are a few major categories worth supporting by voting for candidates who also support these policies, contacting your representatives to encourage them to support these policies, or maybe even publicly advocating for them yourself.

Clean Energy

First, and perhaps most obvious, is to **support policies that encourage clean or low-carbon energy expansion**, such as through subsidies, grants, deregulation of permitting and installation, or the removal of subsidies for dirtier forms of energy. While making less, using less, and using more efficiently are generally the most sustainable things to do (which I'll cover shortly), from a policy point of view, nothing is as all-encompassing as supporting the expansion of clean-energy sources. Clean energy doesn't just reduce the emissions associated with powering our homes and our things. It also reduces the emissions required to make our things in the first place! This is partially why original estimates on the impact of the Inflation Reduction Act suggest it could drive down America's emissions so considerably.

> Supporting clean-energy policy is one of the most impactful policy positions for the planet.

But what's also important to distinguish is that I did not say "renewable" energy but "clean" energy. Renewable energy generally refers to things that can generate power forever and not run out, like solar energy, wind energy, or hydroelectric energy. But there is another form of low-carbon energy that, while finite, generates the most energy per square foot of installation: nuclear energy.

Nuclear energy has long been demonized, especially (and ironically) by environmentalist groups, for being dirty and dangerous because of questions about the storage of radioactive waste. However, the clean-energy benefits of nuclear energy—including its capability to generate energy on demand, unlike wind or solar—*far* outweigh the waste associated with its production. Nuclear can and should be considered a key piece of our decarbonization strategy, even if it's a bridge until we can more fully transition to true renewable energy.

Here's a great example of two countries' diverging nuclear policies to illustrate this. Over the last several decades, due in part to pressure from environmentalists, Germany has summarily shut down all of its historic nuclear power plants and started shifting investment to renewable energy. France, on the other hand, generates the majority of its electricity from nuclear power.

Electricity Maps is a company I like to reference from time to time that tracks the carbon intensity of various electric grids across the globe. In 2023, about 58 percent of Germany's energy came from renewable sources, compared to only 28 percent of France's energy.[150] But at the same time, Germany has needed to expand its reliance on coal and oil to make up for the lost energy generation from shutting down its nuclear plants. The result is that even though Germany uses almost twice as much renewable energy, its grid emissions intensity in 2023 was about 400 grams of carbon dioxide per kilowatt-hour of energy generated, compared to 53 grams for France (at the time of this writing). **Nuclear energy has helped France avoid coal—and has meant that its electricity generation is several times cleaner than Germany's, despite Germany's grid having more than twice as much renewable energy.**

Discussion Points

1) What reasons did environmentalists originally have for discouraging nuclear power generation? Why is that stance changing today?

2) What is preventing people and companies from installing solar panels on their homes? How could policy changes help make this process easier?

Electrification

Assuming we are eventually successful in shifting our electric grids toward being powered by more renewable energy, the next set of policies to sup-

---

150 "Climate Impact by Area," Electricity Maps, accessed July 2024, https://app.electricitymaps.com/map.

port is to **accelerate the electrification of our machines, vehicles, and homes**.

But first, what do I mean by *electrification*? We may at first assume that most everything we do that uses energy is powered by electricity, but the reality is that there are many types of things we do every day that generate power through some other mechanism.

For example, if you are heating your home or taking a hot shower, you may be burning gas—not using electricity—in order to generate that heat. If you own a gas-powered vehicle, you are burning gas to move your car. If you are eating a burger, the cow it came from generated a lot of methane gas. And for the things you buy, sometimes there is so much heat needed in the manufacturing process (to melt or weld things, for example) that it's more viable to burn fuels than to try to generate that amount of heat with electricity.

> Things that can be powered by electricity have greater potential to decarbonize than things that can only be powered by gas.

At its core, **the more things can be powered by electricity, the more things can be powered by clean energy**. It would almost seem silly to install all this clean-energy infrastructure only to not use it and keep burning gas instead! This is another reason that the Inflation Reduction Act of 2022 was such a great piece of climate legislation: It built in subsidies to accelerate the process of electrification in two key areas, transportation (via EV infrastructure) and heating (via electric heat pumps).

Policies that help strengthen our electric grids are also a key piece of the puzzle. There certainly is a risk of rolling brownouts or blackouts if we electrify too fast without making sure our electric grids can keep up with the added demand. For example, one common cause of brownouts globally is heat waves, as the surge in electricity demand to run AC can sometimes exceed the grid's capability to deliver. Policies to modernize and expand our grids can help with tackling some of the baked-in effects of climate change as well as our solutions to combat it.

> **Discussion Points**
>
> 1) What are some places in your life where you aren't using electricity today? Would these benefit from being electrified? Why or why not?
>
> 2) As climate change increases the likelihood of heat waves and major storms, it also increases the risk of power outages. How should we balance the need for climate resilience with the need for climate action?

Energy Efficiency and Improved Utilization

The next policy position to consider is tied to the notion of what we can do to make the transition to a fully electrified, clean-energy economy easier. One major answer—and one of the biggest levers we've talked about so far in this book—is to **support policies that mandate more stringent efficiency standards**, both in terms of how much energy we use and in terms of how much virgin material we use. I'll start with energy, which is more straightforward, before talking about material efficiency.

Machines (and software) are not built equally. Depending on how efficient they are, they may require more or less energy to complete the same task. Two similar cars may have different fuel efficiency and miles per gallon, for example. An LED light bulb will need far less energy than an old-school incandescent light bulb. Even in the software space, if something makes your computer overheat and the fans run really loudly, chances are it is a piece of software that is not optimized well and is drawing more energy from your computer than it needs to.

Unfortunately, while in some cases being more energy efficient is a competitive advantage (i.e., a more efficient refrigerator may lower your electric bill), oftentimes the relative advantage is a small fraction of the overall cost of ownership of the machine. In many cases, the manufacturers of these goods or software don't have a particularly powerful incentive to make their products function as efficiently as possible. Just look at how much *bigger* cars in America have gotten over the last several decades, despite smaller cars needing less energy. Energy efficiency is not a given in a free market economy.

From a policy point of view, it is valuable to support any sort of energy efficiency standards being proposed. This includes things like miles per gallon requirements for all new cars*, updates to product labels like Energy Star to expand the minimum efficiency limit or the types of machines under its jurisdiction, or regulation of energy-wasteful activities like proof-of-work cryptocurrency mining. The net benefit is that, if all our things become more efficient, we'll need less energy overall—meaning that we don't need to build as many solar farms or wind turbines to get to a clean-energy economy. But here's the flip side: If, for example, our machines get 10 percent *less* efficient, then we'd need to install 10 percent more clean energy *just to offset the new electric demand and get back to where we started.* We have to do these things in parallel.

It is also critical to make sure these standards don't have loopholes. (In)famously, the "Corporate Average Fuel Economy" efficiency standards originally set in 1975 have led to worse fuel economy overall due to a carveout for "light trucks." The unintended consequence was for cars to keep getting bigger rather than more efficient to meet the definition of "light truck" and avoid these efficiency standards. Most cars sold in the US today, including SUVs and crossovers, are classified as light trucks as defined by this regulation.[151]

> More population density helps to better distribute resources and enable lower-carbon forms of transportation.

An important corollary to energy efficiency is to support structures that are, at their core, far more efficient than their counterparts. Thinking back to some of what we covered in part 1, there are two very powerful but very simple high-impact ways to do this: **support policy that increases population density, and support policy that increases public transit**. These often go hand in hand, as increased population density can lead to the rise of more effective public transit.

I'll talk more about the housing policy side of this later (as there is much more to say here beyond just energy efficiency). For now, think back to some of the points made in part 1. Buildings that can fit a lot more people into them (e.g., apartment complexes) and maximize the ratio of people to square feet are generally more effective at distributing energy and resources to those who live in them. Imagine the opposite, where everyone lives in their own mansion.

---

151 Kate S. Whitefoot and Steven J. Skerlos, "Design Incentives to Increase Vehicle Size Created from the U.S. Footprint-Based Fuel Economy Standard," *Energy Policy* 41 (2012), https://doi.org/10.1016/j.enpol.2011.10.062.

WHAT WE CAN DO

Think about all the wasted electricity that would come from lighting, cooling, building, and heating those empty rooms! Tall apartment complexes are a more efficient way of housing people than other forms of housing, and in general, density and better utilization of shared resources make urban living the most sustainable version of living. (And on the corollary, it makes suburban living one of the least sustainable versions.)

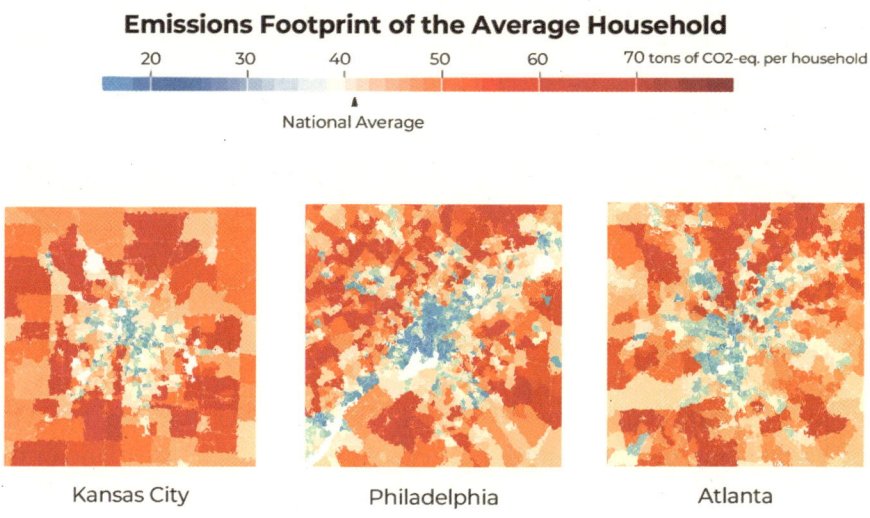

Source: "The Climate Impact of Your Neighborhood, Mapped," *New York Times* (online), https://www.nytimes.com/interactive/2022/12/13/climate/climate-footprint-map-neighborhood.html.

As for public transit, like I said before, the more we can maximize passengers per vehicle, the better. Unless everyone starts carpooling on every drive, buses, trains, subways, and infrastructure for smaller vehicles (e.g., bikes) will always be more efficient ways to move people. Policies to expand alternative transit options (and incentivize their ridership) will be better for the planet—and for traffic! As an example, check out this poster from the city of Münster in Germany that visualizes this well by showing how much space is needed to move the same number of people via bikes, cars, or a bus.

Source: "1991—Münster Congestion Visualization," Data Physicalization Wiki, http://dataphys.org/list/congestion-visualization/.

In addition, it helps to embrace "pedestrian-first" design when voting on city policy. Things that make a city more walkable, such as (but not limited to) better sidewalks, mixed-use zoning (so people can shop closer to where they live), and eliminating "stroads" (two-way streets with a shared turning lane in the center), can also enable denser living and more vibrant city and town centers.

Discussion Points

1) What are the potential risks of focusing only on clean energy and not on energy efficiency?

2) Why do you think we are so dependent on our cars? What do you think could help us become less car-dependent? What would it take to do so?

Material Efficiency and the Circular Economy

When talking about material efficiency, you may have also heard the term **circular economy** to describe the phenomenon. At its core, this is a position of **keeping our materials in use for longer before they are discarded or used up**. It focuses on four things, each with its own policy potential:

1) Use as few resources as possible to build our things.

WHAT WE CAN DO

2) Use as many recycled or reused materials as possible to build our things.

3) Design our things to work as long as possible.

4) Reuse, refurbish, or recycle those things instead of throwing them away.

Why is this so important? As we've seen many times already in our "Make It, Move It, Use It, Lose It" framework, the sheer amount of waste, water, and emissions needed to make and move things often exceed the emissions, water, and waste associated with using things. So anything we can do to not have to make new things in the first place (or mine new raw materials) helps to massively reduce the waste of making things.

Use as Few Resources as Possible

When it comes to using as few resources as possible to build our things, I'd like to extend this to **using as few resources as possible to deliver the same service**. There are some things that just take *way* less to deliver the same goods, such as meat alternatives, as we explored in part 1. But there is often a lot of unnecessary waste in making things and delivering services in the first place.

Let's talk about grocery stores. As we talked about previously, the USDA estimates that between 30 and 40 percent of all food in America is thrown away. That's over 100 *billion* tons of food grown laboriously by our farmers that just become trash. It is staggering.

There are many reasons why this is, and it is not just the fault of grocery stores. But generally, **we as a society have prioritized "availability of everything" over "utilization of everything."** A store would far rather buy ten watermelons, sell seven, and throw out three than buy six watermelons, sell them all, and irk the seventh buyer, who might take their money elsewhere to get their watermelon.

Unfortunately, there aren't a lot of clear-cut policies in the public sphere that immediately tackle this issue. But ones that put disincentives to companies for overproducing and throwing away excess—such as a waste tax or limits on the amount of waste generated overall—could encourage companies to better right-size their supply to customer demand.

Use as Many Recycled or Reused Materials as Possible

This is critical for two reasons. First, we can avoid the waste, water, and emissions associated with needing to mine new raw materials for everything we make (ideally at a rate greater than the emissions associated with recycling). Second, it creates more demand for recycling infrastructure and structures to feed old materials back into new products. Though this does not always lead perfectly to lower emissions (like we saw with the LEGO example in chapter 10), **supporting policies that require a minimum number of recycled, reused, or reusable materials in products can, en masse, enable a more circular economy**.

However, not every type of material is equal when it comes to recycling. Metals like tin, aluminum, and gold are considered "infinitely recyclable," which basically means that they can be reformed into new things without loss of structural integrity. Plastics, on the other hand, tend to degrade every single time they're recycled. So in this case, policies that reduce the amount of single-use plastic (like plastic bag bans you have likely encountered in some stores) in favor of more durable, reusable materials summarily help shift us away from a material that has an end date to how many times it can be recycled.

Design Our Things to Work as Long as Possible

One of the best bipartisan, environmentally friendly policy recommendations that has come up in the last decade is Right to Repair. In other words, the users of a machine should have a right to repair their devices. Whether the device is farm equipment, your phone, or your kitchen appliances, making sure that companies don't artificially prevent you from fixing your things can help you keep your things working for longer (and get you more value out of your stuff at the same time!). Remember, the longer that you can use your thing, the longer you can avoid needing to spend all the energy, water, and waste needed to make a new thing.

Fixing our things rather than replacing them is generally a net benefit for the planet.

Part and parcel of extending the lifespan of products is ensuring robust secondary markets for our things. As we talked about in part 1, buying refurbished or reused things rather than new things is better for the planet. Policy-wise, **anything that creates better access to the availability of used or refurbished products is worth supporting**. Things like planned obsolescence, which artificially truncates the lifespan of a product, can prevent secondary markets from taking off. (Imagine if there weren't a used car market because auto manufacturers had designed a "kill switch" for the engine of a car after it passed 50,000 miles!)

Reuse, Refurbish, or Recycle Things Instead of Throwing Them Away

It is now time to talk about the state of the recycling industry. In a fascinating *Time* article from 2016 entitled "The History of Recycling in America Is More Complicated Than You May Think," it outlines how the origins of recycling are illustrative of why we are where we are today.[152] The article rightly points

---

152 Olivia B. Waxman, "The History of Recycling in America Is More Complicated Than You May Think," *Time* (online), November 15, 2016, https://time.com/4568234/history-origins-recycling/.

out that recycling and reusing have existed since time immemorial, especially when materials and things were in short supply. Humanity often had no other choice.

Source: "The History of Recycling in America Is More Complicated Than You May Think," *Time* (online), https://time.com/4568234/history-origins-recycling/. This poster was originally printed during World War II.

However, during its post–World War II growth phase, America started earnestly down the pathway to its obsession with single-use and disposable items. This led to a massive waste problem that had never existed before—and people were looking for who needed to fix the problem *and* who was to blame for it.

In the early 1950s, a group of entities, including manufacturers responsible for designing these disposable items, got together and founded a nonprofit you may have heard of: Keep America Beautiful. This nonprofit was responsible for one of the most widely known anti-litter ads of all time from the '70s, featuring a person dressed as a Native American man, crying a single tear as a narrator solemnly states, "People start pollution. People can stop it."

Herein lies the problem: **The burden for solving litter was very intentionally and specifically placed upon "people" for recycling, not the manufacturers responsible for making the waste in the first place**.* As part of this "shifting the blame" from producers to consumers, taxpayer money went less into curtailing the spread of waste from producers and instead to the build-out of municipal recycling programs. As the *Time* article very astutely puts it, "in short, recycling stopped being a way for consumers to get more from their purchases and became something that cost people money or at least time."[153]

Unfortunately, this marketing strategy is not limited to recycling. The term "carbon footprint" was popularized by British Petroleum (BP) in 2004 to shift the onus on reducing emissions to individuals rather than itself.[154] Thankfully, in the two decades since, regulatory scrutiny has shifted back toward the carbon footprint of companies—including BP.

Fast-forward to today: Recycling has gotten incredibly complex for the average person. There are seven different "types" of plastics with recycling logos (if the logos are visible at all), and which types are curbside-recyclable can vary from municipality to municipality. It's not always fully clear what condition the plastic has to be recycled in (e.g., if there's a bit of food left on it). And worse, there are many "seemingly recyclable" things that are secretly nonrecyclable, like certain takeout containers or drink cartons.

As a result, even when we do our best to recycle our things correctly, so many people "flood" recycling with nonrecyclable items that a significant majority of things labeled as recyclable just get thrown out anyway by recyclers. In 2018, only 9 percent of all plastic was actually recycled, per the EPA.[155] This

---

153 Waxman, "The History of Recycling."

154 Julie Doyle, "Where Has All the Oil Gone? BP Branding and the Discursive Elimination of Climate Change Risk," *Culture, Environment and Ecopolitics* (2011), https://www.academia.edu/4030712/_Where_has_all_the_oil_gone_BP_branding_and_the_discursive_elimination_of_climate_change_risk

155 "Frequent Questions Regarding EPA's Facts and Figures about Materials, Waste, and Recycling," United States Environmental Protection Agency, last modified November 22, 2023, https://www.epa.gov/facts-and-figures-about-materials-waste-and-recycling/frequent-questions-regarding-epas-facts-and.

number may have dropped to as low as 5 percent in recent years after China (historically America's primary purchaser of recycled plastic) banned many "waste imports" after 2018.[156]

Other "recyclable" items didn't fare much better. Only 17 percent of aluminum and 25 percent of glass were recycled. Only paper and paperboard (including packaging) fared relatively well, at 68 percent.[157] The financial ROI for many recyclers has materially diminished, to the point where California passed a Truth in Labeling law preventing the application of a recyclability logo to items that are generally not recycled.

So what should we do about making recycling better from a policy perspective? Generally, recycling should be a "last resort." As mentioned earlier, anything that supports more reusability of items or mandates the use of recycled materials can help drive more ROI for recycling infrastructure. However, there is an emerging policy concept to keep an eye out for that's called **Extended Producer Responsibility**. In short, it puts the cost of the waste *back to the producers of the waste* rather than to the consumers of the goods—reversing some of the decisions made back in the '50s. Generally, a cost incurred by a company but paid for by someone else is considered an **externality**. A great example of this is Coca-Cola making single-use cans and bottles but having municipalities use our tax dollars to clean and recycle them. If the true cost of recycling had to be borne by the companies producing the waste, you'd better believe that they'd be incentivized to design goods and packaging with far more recyclable or reusable parts.

Until then, the best advice on recycling we should follow is, sadly, "when in doubt, throw it out."*

*There are also emerging categories of goods—especially plastics—that are classified as biodegradable or compostable. On the surface, these seem fine to throw out. If you read the fine print, however, these plastics typically require industrial composting facilities to decompose; if they are discarded incorrectly, they may end up in a landfill without the right conditions to decompose effectively.

---

156 Emily Barone, "U.S. Plastic Recycling Rates Are Even Worse Than We Thought", *TIME* (online), May 19, 2022, https://time.com/6178386/plastic-recycling-rates-overestimated/.

157 "Advancing Sustainable Materials Management: 2018 Fact Sheet", United States Environmental Protection Agency, last modified December 2020, https://www.epa.gov/sites/default/files/2021-01/documents/2018_ff_fact_sheet_dec_2020_fnl_508.pdf.

> **Discussion Points**
>
> 1) When was the last time you repaired something? Was it easy or hard? Why?
>
> 2) Recycling is often perceived as one of the more sustainable things that people can do in their lives. What has led to this perception? What could change that?
>
> 3) What might it take for the "Reduce" and "Reuse" parts of "Reduce, Reuse, Recycle" to be considered first for an average American rather than defaulting to "Recycle"?
>
> 4) The recycling logo makes it seem like all three parts of "Reduce, Reuse, Recycle" have equal impact. Sketch a new "recycling" logo that highlights a more accurate hierarchy.

Market Incentives

Now that we've mentioned extended producer responsibility, it's pertinent to talk about another great set of policies to support: sustainable market incentives. I'll talk about three types of policies that provide more incentives for the market to "self-regulate" and operate more sustainably:

1) Carbon taxing

2) Mandatory climate risk and emissions reporting

3) Anti-greenwashing enforcement

Supporting policies that institute any of the above is an excellent way to force companies to "figure out how to become more sustainable" rather quickly, without necessarily mandating specific actions they need to take to do so.

Carbon Taxing

This was originally an idea born and popularized by conservative economists and championed by Nobel-Prize-winning economist Milton Friedman. The notion is simple: Rather than ban certain non-sustainable activity, why not charge extra for it? The reason why this was originally a conservative policy was that it supposed that the free market, not government intervention, would "take care" of emissions and waste. In their natural course of business, companies would look to minimize their carbon tax burden by seeking lower-cost (i.e., sustainable) alternatives for products and production.

Interestingly, Microsoft—and several other companies around the world—has instituted a carbon tax for its operations. For our company, each business group is taxed based on its emissions at an increasing rate and aims to index to the external cost of offsetting those emissions. The "proceeds" of this internal carbon tax are spent by Microsoft to invest in carbon removal or mitigation to apply to our operations and products. In other words, we have aimed to bring our externalities back to being an internal cost!

What I have experienced is a compelling behavior shift within my company as a result. Instead of sustainability being some "other thing to do," it has started to affect the margin of our products and become embedded in the standard business operating model. Executives and their finance leads will see a line item for a product's carbon tax, then ask us to investigate ways to minimize that tax, such as through lower-carbon, more-energy-efficient product design or partnering with our suppliers to install clean energy in the production of our products. The fee doesn't necessarily dictate "how" to decarbonize, but it does create the financial incentive for leaders to invest more in sustainability, whether they care much about it or not.

The tricky piece about carbon tax policies is to consider three things:

1) **Is the tax big enough to generate sufficient change in the market?** If it's a rounding error, companies may very well ignore it and continue to pollute, especially if the marginal benefit of polluting more exceeds the added cost.

2) **Is the tax putting the cost in the right place?** Another concern is that corporations will pass the added cost directly to consumers rather than aim to decarbonize themselves or accept lower margins. This is often a criticism levied against a carbon tax on gasoline, for example, as it can raise the price of gas without materially changing the behavior of the companies producing the gasoline. This concern is especially pertinent when there aren't competitive market dynamics that could reward a company being taxed for pursuing less-polluting solutions.

3) **Are the tax proceeds being used to drive sustainable outcomes?** Ideally, the money generated from a carbon tax should go

directly to investments that could help lower that tax in the future (e.g., through investment in clean energy or public transit). However, if the carbon tax just becomes a slush fund for the government, it could actually do more harm than good depending on how the proceeds are spent.

There are entire books by very intelligent people dedicated to discussing carbon tax policy. I encourage you to explore them further if this topic is of interest to you.

Mandatory Climate Risk and Emissions Reporting

If carbon pricing is meant to impact a company's cost structure, then this second market incentive is more about impacting a company's fundraising and revenue opportunities. Thinking back to ESG investing, which was introduced in chapter 11, one of the critical pieces of information that investors will need to know about is climate resilience. For example, would you invest in a condominium in Florida if you knew that its specific location was particularly susceptible to flooding and hurricanes, each of which is made worse by the climate crisis? Would you feel comfortable investing in a home insurance company that covers this area if you didn't know about those risks?

Every publicly traded company in America is required to disclose material financial risks that could impact its capability to operate and remain profitable. Climate risk is just one additional type of risk that must be understood for investors to make informed choices as to what to invest in. While this should be noncontroversial, this is an active political battle. For example, Florida passed a law in May 2023 that aimed to remove ESG considerations from all government and public pension investment decisions.[158]

Beyond just reporting the risks of climate change, there is also value in supporting policy (such as that explored by the Federal Trade Commission during the Biden administration) to require disclosure by companies as to their emissions as part of their financial reporting. Not only does this put increased pressure on public companies to make sustainability commitments, but it also lets investors make more informed choices as to which companies are materially reducing their emissions versus which companies are not taking sustainability as seriously. And by mandating these disclosures, this becomes a new measurable metric that business CEOs and CFOs can track and ideally aim to

---

158 "DeSantis Signs Sweeping Anti-ESG Legislation in Florida," Reuters, May 3, 2023, https://www.reuters.com/business/sustainable-business/desantis-signs-sweeping-anti-esg-legislation-florida-2023-05-02/

mitigate so that they can appear more attractive to investors and customers of their products.

Anti-Greenwashing Enforcement

Greenwashing is a term that generally refers to attempts by organizations to appear more sustainable than they are through misleading marketing and advertising. In some regards, the recycling discussion from earlier is a form of greenwashing. We, as consumers, are led to believe that our products are recycled if we put them in the green bin, when in reality, most are not. You've probably seen brand upon brand in the store claiming to be "eco-friendly" or "sustainable" in one way or another. But without strong anti-greenwashing claim enforcement, you could be misled!

Generally, policies that look to standardize measurement and definition of sustainability to create a common basis for comparison, like ecolabeling (e.g., Energy Star for energy efficiency, LEED certifications for building efficiency, fuel economy for cars), are a great start. However, it's complex, given the tricky nature of measuring "true" impact, as we've covered throughout this book. But that, at least, helps fight the confusion that could stem from one company claiming they are sustainable because they used bio-based plastic, another company claiming it because they have recyclable packaging, and a third one claiming it because they bought carbon offsets. These three things are completely different, and many companies will claim "sustainability" because they did the easiest (and often cheapest) "sustainable-sounding" thing rather than taking a truly comprehensive approach to making their product more sustainable. Ecolabels help take the guesswork out of the equation for both customers and producers—a win for both parties if done correctly.

Discussion Points

1) Do you believe that market incentives can reasonably lead to more sustainable outcomes? What could get in the way of realizing those outcomes?

2) Have you ever experienced an example of greenwashing? What tricks did the organization use to overstate its environmental focus?

Environmental Protection

While there are certainly more policies we could talk about, I'll wrap up this discussion on "obvious" policies to support by talking about one of the most publicly popular kinds: the conservation of existing land, water, and wild animals.

Biodiversity is one of the most critical things keeping our planet alive and thriving. And yet without proper protection in place, it can become seriously threatened. Thankfully, we've been pretty good at passing landmark policies to protect many of our at-danger ecosystems, such as the Clean Water Act and the Endangered Species Act of the early 1970s (as part of Richard Nixon's administration) and the creation of national parks and forests in the late 1800s and early 1900s (as part of Theodore Roosevelt's administration).

Beyond our love for animals and nature, why are these laws so important for climate change? While the answer is incredibly complex, it can be boiled down to two major facets: They make our environment more resilient, and they capture greenhouse gas emissions while slowing down climate change. (And, perhaps, another benefit is that the beauty of nature has inspired you to want to protect these lands and our planet.)

The resiliency piece can be seen in a few ways. Per the Federal Emergency Management Agency, natural floodplains like wetlands are incredibly effective ways to mitigate the risk and damages of flooding and erosion while being a friendlier environment for wildlife.[159] When we destroy those floodplains and replace them with things like asphalt (which doesn't absorb water), the flood risk can be exacerbated. The city of Houston, which is famous for its car-first infrastructure and the widest freeway in the world at twenty-six(!) lanes, suffered from significant flooding after Hurricane Harvey, due in part to its removal of natural floodplains in favor of asphalt and cement.

Another way that conservation can improve resiliency to the effects of climate change can be seen in how biodiversity creates natural redundancies. Per the USGS, "reduced species or genetic diversity . . . could lead to a reduced capacity for ecosystems to respond to additional stresses."[160] Imagine a not-so-biodiverse food chain that is reliant on one single type of bee rather than on many different types of pollinators. If something were to happen to that bee population (e.g., illness, an invasive species, climate stressors), then all

---

[159] "Benefits of Natural Floodplains," FEMA, last modified April 1, 2022, https://www.fema.gov/floodplain-management/wildlife-conservation/benefits-natural.

[160] "Biodiversity Critical to Maintaining Healthy Ecosystems," United States Geological Survey's Wetland and Aquatic Research Center, January 15, 2016, https://www.usgs.gov/news/biodiversity-critical-maintaining-healthy-ecosystems.

of a sudden this food chain would collapse. Plants wouldn't get pollinated and would die off, and all the animals that eat those plants would starve.

There's a very real example of this with sea otters and kelp off the US West Coast during the early twentieth century. Due to the overhunting of sea otters, people witnessed a significant (and surprising) decline in kelp forests. Why? Sea otters were the only main predator of sea urchins, which fed on the kelp. Without a predator to keep them in check, urchins would decimate kelp forests, creating "urchin barrens" that prevented kelp from returning—and harmed other species that were reliant on kelp, like shrimp. Things we can do to protect biodiversity will build resiliency into nature so it can withstand these stressor events.

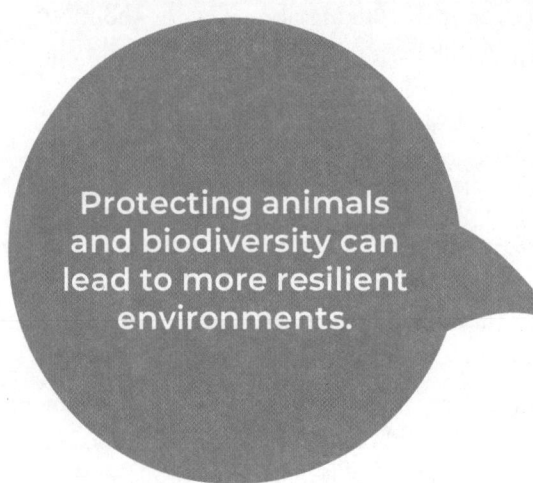

Protecting animals and biodiversity can lead to more resilient environments.

This is especially important when considering that nature is a really important carbon sink for climate change. In its annual *Inventory of U.S. Greenhouse Gas Emissions and Sinks*, the EPA tracks the total amount of carbon sequestered (i.e., removed from the atmosphere) by nature. In 2022, nearly 800 million metric tons of greenhouse gas emissions were absorbed by "forest land remaining forest land" (i.e., not being cut down), with an additional 100 million tons from "land converted to forest land" and 140 million tons from "settlement" trees (i.e., urban, suburban, and rural trees). Altogether, trees absorb over 15 percent of all US emissions annually and store far more than this in their mighty trunks.[161]

Sadly, in forest fires—a result of climate change made worse by development, improper management, and biodiversity loss—trees emit a lot of greenhouse gas emissions back into the atmosphere. So though American trees absorbed over 900 million metric tons of emissions in 2022, they also emitted over 100 million tons back into the atmosphere from burning down—twice as bad as only three years prior in 2019. To visualize how much forest burned

---

161 United States Environmental Protection Agency, *Inventory of U.S. Greenhouse Gas Emissions and Sinks*.

down in 2022, this is the equivalent of over 7 million acres of forest—larger than the entire state of Massachusetts.¹⁶²

So, **policies that protect biodiversity and conserve land, water, and forests are some of the best protectors of our natural capability to fight climate change**. It's critical to protect not just the wildlife and plant life but also their sources of food and water, their predators (to keep their population in a healthy balance), their habitats, and the climate they thrive in. The best policies are the ones that are the most holistic; point solutions, like introducing a predator into an ecosystem to fight an invasive species, can often backfire.

But forests and their denizens alone are not the only carbon sinks that we should care about and prioritize. Grasslands also absorb greenhouse gas emissions from the atmosphere, as does farmland. Regenerative agriculture, a relatively new term in the climate discourse, is a method to shift our farming practices to regenerate the soil we till, absorbing more greenhouse gases into our soils in the process (which is healthy for the soil!).

> Planting trees and soil-regenerating, diverse plants can absorb emissions from the atmosphere.

Today, farmland sequesters about 30 million tons of greenhouse gases annually, despite the vast majority of agriculture being classified as "non-regenerative" today. One could easily see the amount of carbon removal from farmland grow significantly as a result of more widespread regenerative agriculture techniques, helping build more robust soil for our food in the process. While formal estimates for impact are hard to find, more research is being done in this space by both the private and public sectors.

There is also a very large source of carbon sequestration that is left out of America's annual greenhouse gas reporting: water! Though we don't count ocean carbon sequestration because America doesn't "own" any oceans, oceans absorb roughly a quarter of all greenhouse gas emissions and 90 percent of the excess heat generated by our emissions, per the United Nations.¹⁶³

---

162 "Annual 2022 Wildfires Report," National Oceanic and Atmospheric Administration's National Centers for Environmental Information, January 2023, https://www.ncei.noaa.gov/access/monitoring/monthly-report/fire/202213

163 "The Ocean—the World's Greatest Ally Against Climate Change," United Nations, Climate Action (blog), accessed July 2024, https://www.un.org/en/climatechange/science/climate-issues/ocean.

And even if we don't "count" them in our numbers, we nevertheless engage with oceans and coastal habitats. Per the National Oceanic and Atmospheric Administration, the term "blue carbon" refers to carbon dioxide absorbed by our oceans, seagrasses, mangroves, coral reefs, and other coastal ecosystems. While forests take a long time to grow, these "blue carbon" solutions can absorb and store carbon at a much faster rate than forests—and can often be a highly valuable ecosystem to protect and expand from a policy point of view.[164]

Before moving on to our next set of policies, I want to address one of the more trendy carbon sinks in modern discourse: **peatlands, wetlands, marshes, and bogs**. Advocates for these watery land features point most frequently to their capability to store carbon dioxide like nothing else. The United Nations estimates that wetlands and the like actually sequester *twice* as much carbon dioxide as all of the world's forests combined.[165]

However, the trickiest thing about climate change is that there are *many* types of emissions that cause it. Carbon dioxide is one, but it is also among the least potent: Methane is 28 times worse than carbon dioxide, nitrous oxide is 265 times worse, and certain hydrofluorocarbons can be thousands of times worse. (Hydrofluorocarbons, also known as HFCs, are ironically used to replace ozone-depleting chlorofluorocarbons [CFCs], the dangerous chemicals historically used in refrigerators and aerosols).[166] Methane production, in particular, is what complicates the picture on wetlands.

According to the EPA's *Inventory of U.S. Greenhouse Gas Emissions and Sinks*, natural coastal wetlands sequestered 11 million metric tons of carbon dioxide but produced 4 million metric tons of methane, netting a smaller amount (7 million metric tons) of emissions sequestered. If you factor in the creation of artificial wetlands—such as reservoirs, canals, and ponds, which the EPA denotes as "flooded lands" and counts as wetlands—these *added* about 44 million metric tons of emissions. These are almost exclusively due to methane, especially in Texas and Florida, due to their climates.[167]

Is that to say we shouldn't support wetlands (both natural and artificial) as ways to combat climate change? Not at all. Natural wetlands are critical habitats that build resilience against floods and have a net benefit to emissions, and artificial wetlands (specifically reservoirs) are often critical sources of drinking

---

164 "What Is the Carbon Cycle?," National Oceanic and Atmospheric Administration's National Ocean Service, accessed July 2024, https://oceanservice.noaa.gov/facts/carbon-cycle.html.

165 "Peatlands Store Twice as Much Carbon as All the World's Forests," United Nations Environment Programme, February 1, 2019, https://www.unep.org/news-and-stories/story/peatlands-store-twice-much-carbon-all-worlds-forests.

166 United States Environmental Protection Agency, *Inventory of U.S. Greenhouse Gas Emissions and Sinks*.

167 United States Environmental Protection Agency, *Inventory of U.S. Greenhouse Gas Emissions and Sinks*.

WHAT WE CAN DO

water for townships or sources of carbon-free hydroelectric power. But things like cosmetic artificial ponds? Maybe we can pull back a little bit on those.

Discussion Points

1) Have you witnessed threats to wildlife or biodiversity? What is leading to those threats? How could they be addressed?

2) Planting trees is often perceived as one of the top sustainability actions we can take. What has led to that perception? When could planting trees create problems?

Chapter 15: Sustainability-Adjacent Policies to Support

Once you move past policies that are explicitly about climate change and environmental protection, what are the next types of policies that are worth supporting? Which ones may have a significant climate impact without seeming like it? The following is an inexhaustive list of policy categories with a surprising impact on sustainability.

Accessible Family Planning

The first "surprising" set of policy positions that can lead to more sustainable outcomes arises in the form of accessible family planning. Recall from part 1 that decisions around family planning can be one of the most influential decisions we can make in our lives—from whether we adopt to how many children we have to when we decide to have those children.

Today, however, nearly half of families aren't planned at all. Per data from the Centers for Disease Control and Prevention (CDC), 45 percent of all pregnancies in 2011 were unintended.[168] Unfortunately, many of the people who had an unintended pregnancy were in positions that made raising these children harder—specifically, young mothers, mothers living below the poverty line, and mothers who had not completed high school.

Family planning can enable more sustainable family growth.

As we talked about in part 1, *when* we choose to have children can be just as important

---

168 "Unintended Pregnancy," Reproductive Health, Centers for Disease Control, May 15, 2014, https://www.cdc.gov/reproductive-health/hcp/unintended-pregnancy/.

as *how many* children we have when considering climate impact. Helping people understand how to plan a family ends up carrying positive sustainability outcomes as well as other potential societal benefits.

Unfortunately, access to appropriate family-planning resources is not always available—and in some cases, it's explicitly banned or discouraged. (Think, for example, of abstinence-only sex education in favor of more robust sex education.) In some places, access to and promotion of contraception is at risk of significant rollbacks. Places like Planned Parenthood, which advocates for and teaches about reproductive health, have faced the prospect of budget cuts.

A first order of business should be **supporting policies that focus on the "planning" part of family planning**, including more robust sex education, access to contraception (including Plan B), and access to reproductive health resources. That way, parents can choose to stop growing their families at a planned number (rather than having an extra child by accident), have children when they are ready (rather than early), and—in some cases—avoid having children at all if they so wish. Each of these outcomes has a significant benefit to the environment, especially when on a national scale.

The second consideration is much more morally heavy: **options that exist when planning fails**. Abortion—the main way to terminate an unplanned pregnancy—has been one of the most politically fraught topics of the past five decades in the United States, and for good reason. One group of people sees a fetus as a life at conception (and, therefore, they see themselves as preventing the murder of an innocent). The other group sees a fetus as a fundamental part of the mother and her body (and, therefore, they see themselves as protecting the mother's right to choose what to do with her body).

I am not here to weigh in on the moral validity of one position versus the other. That is something that you, the reader, must ultimately do. But from a purely environmental point of view, protecting a woman's right to choose can lead to a more sustainable outcome by lending families another means to planfully start and grow their families.

> **Discussion Points**
>
> 1) Birth rates across the world have fallen considerably over the last century. Why do you think that is? What risks come with lower rates of reproduction?
>
> 2) What do you think contributes to the statistic that nearly half of all pregnancies are unplanned? What types of actions could help reduce this?

Safe and Affordable Cities

Next, let's think back to another of the biggest impacts discussed in part 1: where we choose to live.

We've already concluded that cities and other dense population centers tend to be more climate-friendly places to live due to more efficient sharing of common resources, the need for less infrastructure, and easier opportunities for public transit and walking options, among other things. However, after a decade of consistent growth in urban areas across the world, the COVID-19 pandemic—coupled with the newfound capability to work remotely—began to reverse that trend. (Though it remains to be seen if that reversal is a temporary blip in response to COVID-19).

Per the US Census Bureau, some city populations shrank considerably during the first year of COVID due to migration. Boston's and Washington, DC's, shrank by 2.9 percent, New York City's shrank by 3.5 percent, and San Francisco's shrank by 6.3 percent.[169] And while these cities largely stabilized in the year that followed, they are not replacing the lost residents. For example, the populations of New York City, Chicago, Philadelphia, and Los Angeles all continued to shrink in 2021 and 2022.

It is worth mentioning that certain cities saw a sizable increase in residents during this time. Between July 2021 and July 2022, the top five biggest-growing cities were Fort Worth, Texas; Phoenix, Arizona; San Antonio, Texas; Seattle, Washington; and Charlotte, North Carolina. The growth of these cities did not outpace the number of individuals leaving cities overall, however. On average,

---

169 Amel Toukabri and Crystal Delbé, "Large Cities No Longer the Biggest Population Losers," US Census Bureau, May 18, 2023, https://www.census.gov/library/stories/2023/05/large-cities-no-longer-biggest-population-losers.html.

big cities have shrunk since the start of the decade in favor of suburbs, smaller cities, and more rural areas.

As you can see in the following figure, the biggest relative gains in inhabitants were for towns with a population between thirty thousand and seventy thousand residents. The biggest absolute gains were for cities with a population between seventy thousand and one million residents.

NET DOMESTIC MIGRATION (NDM) BY COUNTY POPULATION SIZE: 2019, 2021, AND 2022

County population size category	2019			2021			2022		
	Population[1]	NDM	DMR[2]	Population[1]	NDM	DMR[2]	Population[1]	NDM	DMR[2]
Less than 30,000...	22,159,950	-24,776	-1.12	21,973,311	125,283	5.70	21,980,903	74,968	3.41
30,000-69,999...	30,128,053	44,611	1.48	30,290,260	200,201	6.61	30,396,710	143,000	4.70
70,000-999,999...	180,033,750	393,861	2.19	183,435,783	770,225	4.20	184,538,306	383,739	2.08
1 million or more...	96,008,200	-384,874	-4.01	96,332,200	-895,264	-9.29	96,371,638	-828,695	-8.60

[1] Sum of the resident population for counties that fall within each population size category.
[2] DMR: Domestic migration rate per 1,000.
Note: Group quarters population change omitted.
Source: U.S. Census Bureau, 2020 and 2022 Vintage Estimates.

Source: "Two Years into Pandemic, Domestic Migration Trends Shifted," US Census Bureau, https://www.census.gov/library/stories/2023/03/domestic-migration-trends-shifted.html.

Which begs the question: In a post-pandemic world, why do people move? Are these moves good or bad for the planet? And are there policy decisions that can help continue to encourage thriving city centers?

Thankfully, the US Census records data on this as well. While the survey questions don't go as deep as one may hope (there is no drop-down answer to say, "I moved to a rural town to buy a larger home so I could work remotely," for example), there are some interesting nuggets to be found in the data.

SPECIFIC REASONS FOR MOVING

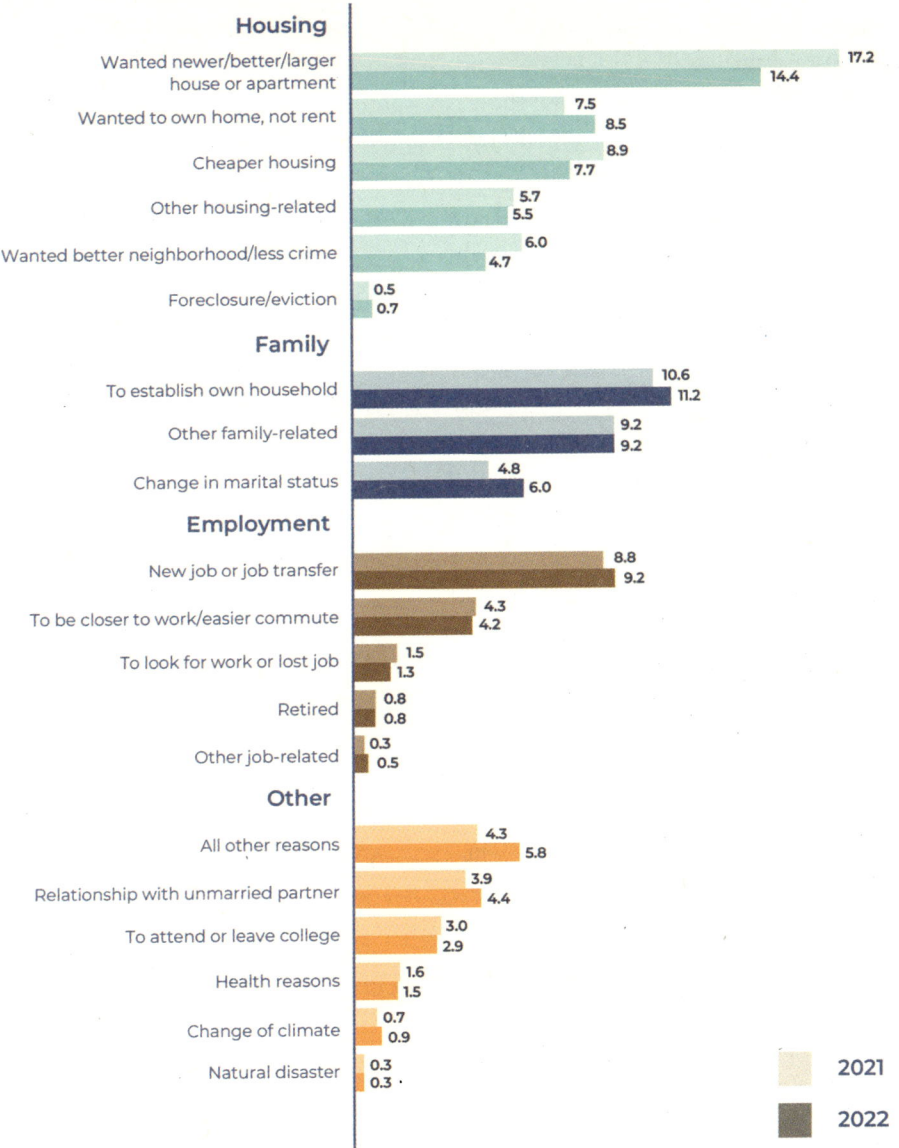

Source: "Why People Move," US Census Bureau, https://www.census.gov/library/stories/2023/09/why-people-move.html.

Based on this information, it seems that some categories for why people move are generally consistent. For example, "I wanted a newer/better/larger house or apartment" is the number-one reason why people move—at least going back to 1998, the earliest year with results from this survey.[170] Family tends to also be one of the most consistent reasons for moving. This was especially true in 2022, when pent-up demand for marriages (and moving in together) following the COVID-19 lockdowns came out in full force, breaking a record for most marriages in a single year since 1984.[171] But if you look specifically at the data from the last few years, there are some interesting trends that couple with a general trend of movement away from cities.

First, in 2021, "I wanted a newer/better/larger house or apartment" was quoted as a reason to move 17.2 percent of the time. This is the highest percentage of respondents pointing to this reason since 2006, right before the Great Recession caused by a significant housing crisis (which was caused, in part, by people receiving bad mortgages to buy homes that were more expensive than they could actually afford). But during a time of COVID-19, remote work, and remote learning, bigger houses became more desirable so that excess space (if available) could be used for things like virtual office or class space. However, as shutdowns went away and people returned to school and work, this desire has also started going down (to 14.4 percent, the lowest since 2008). Instead, "I wanted to own a home, not rent" has grown to 8.5 percent of all moves, the highest it has been since 2006.

Second, "I wanted cheaper housing" grew from 6.7 percent in 2019 and 6.6 percent in 2020 to 8.9 percent in 2021—the highest response rate in nearly a decade—before dropping slightly to 7.7 percent in 2022. People sought out bigger (and cheaper) homes in lower-cost-of-living areas, partially because of historically low interest rates at the time but also because they were empowered by remote working decoupling their job location to their living location. This can also help explain why cities like New York, Washington, DC, and San Francisco—which have much higher costs per square foot of housing—saw an exodus, while cheaper cities and smaller towns saw population gains.

To underscore the effects of remote work in this equation, 2021 and 2022 also saw fewer people indicate that they moved for a new job (only 8.8 and 9.2 percent of respondents, respectively)—the lowest rate since 2011 and down from 12.1 percent in 2019. For the first time, people could reap the benefits of living in a cheaper location while taking advantage of gainful employment in more expensive locations.

Last, in 2021, "I wanted a better neighborhood/less crime" was ranked the

---

170 US Census Bureau, "Table A-5. Reason for Move (Collapsed and Specific Categories): 1999–2022," Current Population Survey, Annual Social and Economic Supplement 1999–2022, July 2024, https://view.officeapps.live.com/op/view.aspx?src=https%3A%2F%2Fwww2.census.gov%2Fprograms-surveys%2Fdemo%2Ftables%2Fgeographic-mobility%2Ftime-series%2Fhistoric%2Fhst_mig_a_5.xlsx.

171 Lorie Konish, "A Record Number of Weddings Are Expected This Year. How to Save if You're Planning a Trip Down the Aisle," CNBC.com, March 18, 2022, https://www.cnbc.com/2022/03/18/how-to-save-on-your-wedding-amid-high-demand-and-inflation.html.

highest it has ever been in this survey (going back to 1998) at 6 percent, up from 4.1 percent the year prior. Even in 2022, which saw this number decrease to 4.7 percent, this reason for moving is still historically high compared to prior years and as high as it's been since 2009.

The root causes of this rise are really tricky and delicate to unpack. To start, the rate of crime (both reported and unreported) in America has unequivocally decreased since the 1990s. Per the Federal Bureau of Investigation (FBI), in 1991, the rate of reported violent crime per 100,000 people was 758, and the rate of reported property crime (e.g., burglary) was 5,140 per 100,000 people. In 2020, the rate of violent crime was only 399 per 100,000 people (about 50 percent lower), and the rate of property crime was only 1,958 per 100,000 people (about 62 percent lower).[172]

Yet during 2020, that number, for the first time in three decades, started to rise for violent crimes (even as it continued to drop for property crimes) and generally remains above pre-pandemic levels at the time of this writing. So even though we are still in an era of exceptionally low crime historically, it *has* gotten worse over the last several years. During the same time span, America has also seen drug-related overdoses, led especially by the rise of fentanyl, more than double since 2015. Per the National Institute on Drug Abuse, fentanyl is now responsible for more than twice the deaths of the second deadliest drug to Americans (meth).[173]

Fentanyl has impacted people experiencing homelessness much harder than everybody else. In Los Angeles, for example, the Los Angeles County Department of Public Health reported in 2023 that people experiencing homelessness in 2020 and 2021 were thirty-nine times more likely to die from a drug overdose than other Los Angeles County residents.[174] Unfortunately, this drug epidemic has become very public and visible within cities as a result and is creating a perception (and sometimes reality) of less-safe cities and public services. A 2020 study by the UCLA Institute of Transportation Studies that surveyed over two hundred US transit operators found that 65 percent of people perceived homelessness as having driven down ridership on buses and trains, with about a third of this group considering the effect to have generated a "major

---

[172] "Crime Data Explorer," Federal Bureau of Investigation, May 1, 2023, https://cde.ucr.cjis.gov/LATEST/webapp/#/pages/explorer/crime/crime-trend.

[173] "Drug Overdose Death Rates," National Institutes of Health's National Institute on Drug Abuse, May 14, 2024, https://nida.nih.gov/research-topics/trends-statistics/overdose-death-rates.

[174] County of Los Angeles Public Health, "New Public Health Report Shows Sharp Rise in Mortality Among People Experiencing Homelessness—Increase Driven by Fentanyl-Related Deaths, Traffic Deaths, and Homicides," May 12, 2023, http://publichealth.lacounty.gov/phcommon/public/media/mediapubhpdetail.cfm?prid=4384.

decrease."[175] This, combined with a news apparatus very eager to cover the rise of crime in the wake of the Defund the Police movement in the early 2020s, seems to help explain at least some of the increase in people moving to seek a better neighborhood with less crime.

With all this context being laid out, let's now explore what types of policy positions could help tackle the above three challenges to help cities—which are generally the most climate-friendly way to house humans—stay desirable and livable. How can we address people's top three (nonfamily) reasons for moving: to find bigger/better homes, cheaper homes, and safer homes? While I am not a policy expert myself, there are a few things that seem to support several of these needs at once.

Affordable Cities

The first policy position to support is simple: build more housing. **Any policy that increases the supply of housing in cities is beneficial for the environment in the long run.** Of course, the reality is far more complex, but the underlying math and economics are simple: Increasing a city's housing supply can create new, better homes (if not always as big as what you could find in the suburbs), reduce the cost of homes (due to more choices for buyers, who receive more bargaining power as a result), and, optimistically, fight homelessness by not pricing tenants out of their homes and onto the street.

And this does mean *any* policy. One standout example is zoning restrictions and laws, which are often upheld (such as egregiously in Silicon Valley) to preserve the value of the original neighborhoods. It often imposes size or height limits on new builds, which artificially restrict the capability to introduce things like denser apartment complexes into neighborhoods. But if there is no more land left on which to build or expand, then the effect is to preserve the existing housing supply rather than expand it to meet demand, driving up prices and urban sprawl. This sentiment, often derided as NIMBY-

> Denser housing is a more sustainable alternative than other types of housing.

---

[175] Anastasia Loukaitou-Sideris, Jacob Wasserman, Ryan Caro, and Hao Ding, *Homelessness in Transit Environments Volume I: Findings from a Survey of Public Transit Operators*, Report No.: UC-ITS-2021-13 (Los Angeles, CA: UCLA Institute of Transportation Studies, 2020), https://escholarship.org/uc/item/55d481p8.

ism, "not in my backyard," should be abandoned in favor of YIMBYism: "Yes! In my backyard!" **Any law that removes zoning restrictions on buildings will generally benefit the environment while reducing the cost of homeownership.** We need to build high—and build dense!

An excellent example of this in motion can be found in Minneapolis. In 2018, the city unveiled its "Minneapolis 2040" plan. Among other things, one of the tenets of the plan was that "all Minneapolis residents will be able to afford and access quality housing throughout the city" by 2040.[176] Amazingly, the city council voted twelve to one to eliminate single-family zoning across the entire city, which historically reserved 70 percent of its land for single-family housing. It coupled that decision with a few other "building-friendly" provisions, such as removing off-street parking minimum requirements, which can artificially restrict growth.[177]

Five years later, despite a period of exceptionally high inflation in the early 2020s following the COVID-19 pandemic, **Minneapolis became the first major American city to "beat" inflation** in this time frame. Rent prices between 2017 and 2023 only grew 1 percent, compared to 31 percent growth across America overall, per the Pew Charitable Trusts.[178] Other cities that relaxed zoning laws in that same time frame (e.g., New Rochelle, New York; Portland, Oregon; and Tysons, Virginia) also saw their rent prices stay relatively flat.

---

[176] "Minneapolis 2040—The City's Comprehensive Plan," Minneapolis 2040, accessed July 2024, https://minneapolis2040.com/overview/.

[177] Richard D. Kahlenberg, "How Minneapolis Ended Single-Family Zoning," The Century Foundation, October 14, 2019, https://tcf.org/content/report/minneapolis-ended-single-family-zoning/.

[178] "More Flexible Zoning Helps Contain Rising Rents," Pew Charitable Trusts, https://www.pewtrusts.org/en/research-and-analysis/articles/2023/04/17/more-flexible-zoning-helps-contain-rising-rents.

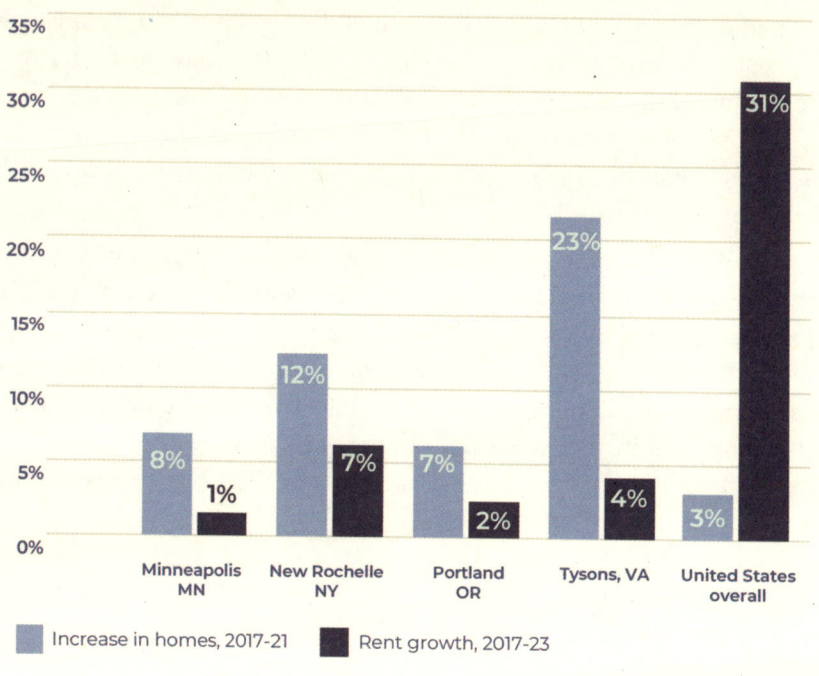

Source: "More Flexible Zoning Helps Contain Rising Rents," Pew Charitable Trusts, https://www.pewtrusts.org/en/research-and-analysis/articles/2023/04/17/more-flexible-zoning-helps-contain-rising-rents.

In a supply-constrained market—which housing is at the time of writing—any new builds, even luxury buildings, have the potential to drive down prices within the market. Even if only the richest people in the city can afford to live in a new building, they are ideally vacating a prior home that could be attainable to the slightly less rich, and so on. As long as the housing supply goes to people who actually live in it, that is.

This is a second point: **Support policy that restricts the capability for housing to be used for nonhousing purposes.** If a city builds one thousand new homes that are purchased by investors who convert them into Airbnbs, foreign investors who never intend to live there and use it to park their money overseas, or snowbirds looking for a getaway home in the winter, then the influx of buildings won't increase the supply of housing in a city.*

*This is why I prefer to stay at hotels rather than Airbnbs when I travel. Hotels are appropriately taxed by cities, purpose-built for short-term stays, and generally incredibly dense compared to Airbnbs.

And depending on an individual city's progress on the above, there is another, perhaps more counterintuitive policy position to consider: **policies that subsidize housing demand can sometimes drive prices up in the long term and *decrease* affordability**. Unless doing so can induce developers to flock to a city to profit from the increased costs, policies that increase demand without increasing supply will reduce the bargaining power of home-seeking individuals who are not actively being subsidized.

This might feel uncomfortable to some people, but programs that are often very well-intentioned to increase access to housing (e.g., rent control) could create unintended knock-on effects in a supply-constrained market to make housing *less* affordable. One of the more nonpartisan think tanks in America, the Brookings Institution, conducted a study on this topic in 2018 authored by a Stanford professor of economics and found that "[w]hile rent control appears to help current tenants in the short run, in the long run it decreases affordability, fuels gentrification, and creates negative spillovers on the surrounding neighborhood."[179] Why? Because subsidies to those who are enjoying rent control increase prices for those who are not enjoying it—and decrease the desire for new housing developers to build new supply (especially if they can go elsewhere to build without restrictions).

Take Saint Paul, Minnesota, the Twin City to Minneapolis right across the Mississippi River. It has historically been the cheaper twin to live in of the two, and in 2021, the city passed what was initially one of the strictest rent-control measures in the nation, capping rent increases at 3 percent for all housing with some room to apply for an exception. Unfortunately, Saint Paul quickly needed to backpedal a bit, as it saw a significant *decrease* in the amount of planned housing development.

Per the United States Department of Housing and Urban Development, the number of building permits issued in Saint Paul for multifamily homes (i.e., duplexes, triplexes, and apartment complexes) *fell* by 45 percent between 2021 and 2022 despite having grown 11 percent the year prior. Minneapolis, which instead prioritized the removal of zoning restrictions as its primary housing affordability policy, saw permits issued for multifamily homes increase by 33

---

179 Rebecca Diamond, "What Does Economic Evidence Tell Us about the Effects of Rent Control?," Brookings Institution, October 18, 2018, https://www.brookings.edu/articles/what-does-economic-evidence-tell-us-about-the-effects-of-rent-control/.

percent over the same period.[180] Whether this initial trend will hold up over the medium and long term remains to be seen, but early data suggests that rent control in Saint Paul (subsidizing demand) has been less effective than zoning reform in Minneapolis (increasing supply) in creating affordable, available housing.

Of course, all of these positions do not have to be all-or-nothing. In some cases, the existence of some form of "housing insurance" limiting evictions or drastic price increases can help prevent a surprise surge of homelessness arising due to a pricing shock or black swan event (e.g., the COVID-19 pandemic). It's generally better if you can couple supply and demand increases together.

And yes, you may recall from earlier that there are notable emissions associated with building new homes, which is completely fair. However, for something that is incredibly supply-constrained like housing, it is fair to assume that, generally, "if the house isn't built here, it'll get built somewhere else." So in this case, while building more city housing does generate near-term emissions, they'll be smaller than the equivalent outside of cities—and they'll drive longer-term environmental benefits (especially if there are rules set forth by cities about meeting efficiency standards for these new builds).

Discussion Points

1) Have you ever considered how housing policy could correlate with sustainability? What are the key ways in which housing and sustainability are interrelated?

2) What are some possible drawbacks to reducing restrictions on home construction? How should these factor into the perspective above?

Safer Cities

Beyond purely focusing on housing policy, there are other policies that are worth supporting to help cities feel safer and more attractive. These also have the potential benefit of driving up the utilization of public transit and walkability, an additional sustainability benefit. (People might be deterred from taking the bus if they don't feel safe on it.)

---

180 "SOCDS Building Permits Database," US Department of Housing and Urban Development, accessed July 2024, https://socds.huduser.gov/permits/.

I will not claim to be an ultimate expert on how to address the third major nonfamily reason causing people to move (i.e., wanting to feel safer or seek a better neighborhood). I am a sustainability expert, not a safety expert, after all. However, there is data to indicate that some of the following are at least good starting points, understanding that public safety is an incredibly intricate and complex topic. And given its political weight, it is hard to find sources truly free of bias. The following list ends up covering the "who's who" of some of the most contentious topics in America. As much as possible, I will aim to remove questions of morality and focus instead on the data.

Where possible, I will try to rely on quantitative, non-editorialized reports from government agencies such as the Bureau of Justice Statistics. For example, it's valuable to start with the framing question of what types of crimes are being prosecuted today. Per said bureau, in 2020, 62.4 percent of all criminals in state prisons were there for violent crimes (e.g., murder, rape, sexual assault, robbery, aggravated assault), 13.5 percent were for property crimes (e.g., burglary, motor vehicle theft, fraud), 12.6 percent were for drug-related charges, and the remaining 10.5 percent were for "public order" crimes (e.g., illegal weapon possession, DUIs).[181]

Now, it is incredibly fair to question whether the number of people incarcerated for the crimes listed above actually correlates to the *amount* of crime happening in each category. For example, it's entirely possible that some types of crimes are caught less frequently or judged more leniently. Also, some crimes may overlap (e.g., someone using illegal drugs commits a violent crime). But by using this as a start, it leads to some interesting takeaways.

Public Health

Surprisingly, a number of studies have found a strong correlation between the expansion of health-care coverage—especially Medicaid, substance-abuse treatment, and mental-health treatment—and the reduction of both drug-based and violence-based arrests.

As it stands today, the correlation between mental health and incarceration is staggeringly high. From a 2016 study by the Bureau of Justice Statistics on the mental health of prisoners, 43 percent of state prisoners have a history of mental-health problems, with 14 percent of all state prisoners having experienced "serious psychological distress" within the thirty days prior to the sur-

---

181 E. Ann Carson, *Prisoners in 2021—Statistical Tables*, NCJ 305125 (Washington, DC: US Department of Justice's Bureau of Justice Statistics, 2022), https://bjs.ojp.gov/sites/g/files/xyckuh236/files/media/document/p21st.pdf.

vey.[182] This has led some to consider the American prison system the country's de facto largest provider of mental-health care in the country, by virtue of services that can be granted to inmates.

One study from Stanford University in 2021 found that losing access to mental-health services increased male incarceration rates by 22 percent compared to men who retained them.[183] Another study by Boston University in 2022 found that counties that expanded Medicaid saw between a 25 to 41 percent reduction in drug-related arrests and a 19 to 29 percent reduction in violence-related arrests.[184]

> Getting therapy, and encouraging others to do so as well, can help reduce emissions.

While there isn't a consensus on this topic yet, given the sheer prominence of health issues behind many crimes and the emergence of new studies linking access to health care with reduced crime, one can extrapolate that both the political (and cultural!) acceptance of expanded access can help to reduce crime, increasing safety and sustainability by extension. That's right—destigmatizing therapy can actually lead to a more sustainable country. Who knew!

Gun Control and Gun Rights

Though gun control is a politically charged policy position, it ends up being relevant to our discussion on crime. The debate, simplified to its impact on crime specifically, comes down to two main perspectives: increased access to guns *increases* crime, and increased access to guns *decreases* crime (by deterring it).

Sadly, after a 1996 amendment known as the Dickey Amendment banned the use of CDC funds to explicitly study gun violence, there is a lack of widely available reputable data to help answer this question. Any conclusions should

---

182 Laura M. Maruschak, Jennifer Bronson, Ph.D., and Mariel Alper, Ph.D., *Survey of Prison Inmates, 2016: Indicators of Mental Health—Problems Reported by Prisoners*, NCJ 252643 (Washington, DC: US Department of Justice's Bureau of Justice Statistics, 2021), https://bjs.ojp.gov/sites/g/files/xyckuh236/files/media/document/imhprpspi16st.pdf.

183 Elisa Jácome, "How Better Access to Mental Health Care Can Reduce Crime," Stanford University's Institute for Economic Policy Research (SIEPR), July 2021, https://siepr.stanford.edu/publications/policy-brief/how-better-access-mental-health-care-can-reduce-crime.

184 Jessica T. Simes and Jaquelyn L. Jahn, "The Consequences of Medicaid Expansion under the Affordable Care Act for Police Arrests," *PLoS ONE* 17, no. 1 (2022), https://doi.org/10.1371/journal.pone.0261512.

be taken with a grain of salt. Plus, at the time of this writing, the formation of a new federal Office of Gun Violence Prevention in 2023 had not yet yielded any research to suggest a conclusion one way or the other.

As a result, there is a bevy of contradictory studies from a broad range of players—sometimes coming to exact opposite conclusions using the same base data. Further, many public articles will make broad-based claims based on decades-old data, including one I found that claimed that concealed carry bans reduced homicide rates based on a study of data *from the 1920s*. (This is an excellent reminder to always check the sources behind any claim you read.)

So rather than cherry-pick a statistic from inconclusive data, I will point instead to one facet of this debate: the expansion of universal background checks for gun purchases, which seems to be universally supported. Even better, this policy position is backed by a reasonable amount of evidence to suggest that it reduces violent crime (or, at the very least, doesn't make it worse). Surveys from the Pew Research Center in 2021[185] and 2023[186] found that nearly 90 percent of both Republican and Democratic votes favored policies that prevented people with mental illnesses from purchasing guns, and 70 percent of Republicans (and 92 percent of Democrats) supported extending background checks to private gun sales and sales at gun shows.

Today, it is already illegal in America to purchase a gun if you are a violent felon or drug addict, are subject to a restraining order, or have a prior conviction for domestic assault. Enforcing laws to prevent these populations from easily obtaining firearms, extending where these laws are enforced, and expanding the category to cover additional at-risk factors (e.g., mental illness) could reduce both real (and/or perceived) criminal activity, leading to the experience of safer and more desirable cities—and therefore encourage denser, more sustainable living.

Policing and Police Funding

We can't talk about public safety without talking about this massively political consideration. Acknowledging that policing and police funding can be a particularly delicate topic for most of us (especially those who have been on the receiving end of injustice), it is at least worth reviewing whether any data exists to correlate policing and reduced crime.

At the bare minimum, cities would likely struggle without at least some

---

[185] Pew Research Center, *Amid a Series of Mass Shootings in the U.S., Gun Policy Remains Deeply Divisive*, April 20, 2021, https://www.pewresearch.org/wp-content/uploads/sites/20/2021/04/PP_2021.04.20_gun-policy_REPORT.pdf.

[186] "2. Americans' Views of Specific Gun Policy Proposals," Pew Research Center, June 28, 2023, https://www.pewresearch.org/politics/2023/06/28/americans-views-of-specific-gun-policy-proposals/.

presence of policing. A 2020 experiment in creating a police-free Seattle neighborhood (formerly known as the Capitol Hill Autonomous Zone or the Capitol Hill Organized Protest) came to a violent and quick end after four shootings ended its one month of existence.

The question is not whether we should have police but rather how much and what kind of policing can reduce crime. Until the symptoms of crime can be addressed and eliminated, policing will be a necessary part of maintaining order in society. Both extremes seem problematic. On one hand, if police are defunded too much (especially without a plan to replace police with other public safety employees), there very well could be an inability by the police to fully protect a city's inhabitants. On the other hand, if police are overfunded, that money needs to come from somewhere—which could be pulling from other crime prevention (or sustainability!) initiatives, addressing the symptom but not the cause of crime.

Unfortunately, like with gun control, there does not seem to be a conclusive answer to this question. Part of the difficulty is that "crime" itself is so varied (e.g., violent crime versus property crime versus white-collar crime). So is the impact of who is affected by crime (e.g., gang violence, crimes concentrated in specific areas of cities). The perception of crime and safety does not always correlate with actual crime rates, either. For example, while Chicago has often been demonized as an unsafe city, its per-capita crime rate is forty-seventh out of America's one hundred largest cities.[187]

Per 2019 data from the FBI, the "least safe" cities in America of the one hundred most populous cities (when including all types of crimes reported) are identified in the following figure, representing both the top ten cities with highest overall crime rates as well as those with highest overall violent crime rates per capita.[188]

To see if there's any high-level pattern, I've overlaid how much each of these cities spends on its police as a percentage of its overall annual budget (as published by the cities),[189] ranked against the same list of the top one hundred most populous cities in America.

Unfortunately, a clear pattern doesn't emerge. My childhood home of Milwaukee, for example, spends a whopping 46 percent of its annual budget on police (the fifth highest percentage among cities) yet still suffers from the sixth worst violent crime rate in the nation. At the same time, San Francisco has the

---

187 "Crime in the United States 2019," Federal Bureau of Investigation's Uniform Crime Reporting Program, accessed 2023, https://ucr.fbi.gov/crime-in-the-u.s/2019/crime-in-the-u.s.-2019/tables/table-8/table-8.xls/view.

188 "Crime in the United States 2019," Federal Bureau of Investigation's Uniform Crime Reporting Program.

189 "City Budgets Belong to Us. How Do America's 300 Biggest Cities Spend Our Tax Dollars?," ACRE Action Center on Race & the Economy, accessed 2023, https://costofpolice.org/.

City	Annual Crime per Capita (Overall)	Annual Crime per Capita (Violent)	Police Funding as % of City Budget
Albuquerque, NM	#1	#9	31%
Memphis, TN	#2	#4	40%
St. Louis, MO	#3	#1	31%
Vancouver, WA	#4	#53	21%
Oakland, CA	#5	#11	43%
Baltimore, MD	#6	#3	28%
San Francisco, CA	#7	#37	10%
Detroit, MI	#8	#2	29%
Baton Rouge, LA	#9	#20	29%
Anchorage, AK	#10	#13	23%
Cleveland, OH	#12	#7	33%
Kansas City, MO	#14	#5	42%
Indianapolis, IN	#21	#10	35%
Milwaukee, WI	#25	#6	46%
Stockton, CA	#29	#8	56%

seventh worst rate of crime in the country while spending nearly the lowest portion of its budget (approximately 10 percent) on its police force.

In other words, there doesn't seem to be a clear link between how much a city spends on its police and how much crime it has. There are, unfortunately, far too many interrelated factors at play to see a clear pattern emerge. It is likely far more important to understand *how* the police are used rather than *how many* police there are. For example, a 2019 Yale University study focused on New Haven, Connecticut, found that community-oriented policing increased trust in law enforcement across the board, no matter the racial or ethnic group of those surveyed.[190] Policies that can engender better community trust in police (e.g., community policing, more oversight of police abuse) are likely a good place to start toward creating safer—and, therefore, more sustainable—cities.

Cleaner and Greener Cities

I remember first reading about the broken windows theory in Malcolm Gladwell's 2000 book *The Tipping Point*. As he wrote back then:

> If a window is broken and left unrepaired, people walking by will conclude that no one cares and no one is in charge. Soon, more windows will be broken, and the sense of anarchy will spread from the building to the street on which it faces, sending a signal that anything goes.[191]

The prime example that Gladwell used to prove this theory was the New York City subway system. At the end of the 1970s, New York's subway was the least safe mass transit network in the world, recording over 250 felonies per week by the end of 1979.[192] Annual ridership dropped by over 20 percent during that decade.[193] The excellent 1979 film *The Warriors* captured a powerful snapshot in time of the subway system's decrepit status, if you are looking for a new movie to watch.

In 1984, the New York City Transit Authority decided to put the broken windows theory into practice with the introduction of the Clean Car Program. Over the next six years, each of the thousands of subway cars was cleaned and graffiti was removed. Any new graffiti was to be removed within two hours(!), or the car would be pulled from service to be cleaned.[194]

The Metropolitan Transportation Authority (MTA) coupled its work of cleaning up the visual element of the subway with the behavioral. Once the

---

190 Kyle Peyton, Michael Sierra-Arévalo, and David G. Rand, "A Field Experiment on Community Policing and Police Legitimacy," *Proceedings of the National Academy of Sciences*, 2019, https://www.pnas.org/doi/10.1073/pnas.1910157116.

191 Malcolm Gladwell, *The Tipping Point: How Little Things Can Make a Big Difference*. New York: Back Bay Books, 2002: 141

192 Mark S. Feinman, "The New York Transit Authority in the 1970s," NYCsubway.org, November 19, 2002, https://www.nycsubway.org/wiki/The_New_York_Transit_Authority_in_the_1970s.

193 "History of the New York City Subway," Wikipedia, accessed July 2024, https://en.wikipedia.org/wiki/History_of_the_New_York_City_Subway.

194 Maryalice Sloan-Howitt and George L. Kelling, "Subway Graffiti in New York City: 'Gettin Up' vs. 'Meanin It and Cleanin It,'" (paper, Northeastern University, College of Criminal Justice, Boston, MA, 1990), https://popcenter.asu.edu/sites/default/files/171-sloan-howitt_kelling-subway_graffiti_in_new_york_city_.pdf.

subways were physically cleaned, the MTA started homing in on other public displays of lawlessness like fare hopping, which was found to be correlated to higher rates of more serious crime. The result was a subway system that both *looked* safer (its broken windows had been repaired) and *felt* safer—because it was! By the end of the 1990s, the rate of felonies had shrunk by around four times compared to the low point of the late 1970s—and ridership increased by almost 40 percent as a result.[195]

Now it is worth asking: Does the broken windows theory prove true in other, interesting ways to reduce crime across America? It seems that the answer is yes! (Or, at least, it doesn't make things worse.)

In 2022, the University of Michigan found that communities that "greened" vacant lots saw reductions in crime. This reinforced a 2018 study that found efforts to maintain vacant lots reduced assaults and violent crime by 40 percent.[196] On the flip side, a study by the United States Forest Service found that the loss of trees in Cincinnati (due to the nefarious emerald ash borer) *increased* crime rates.[197] In general, making neighborhoods seem more warm and inviting tends to decrease crime, and the opposite (e.g., broken windows, vacant lots, excessive litter) could welcome more of it.

> Green spaces help support more sustainable communities.

This leads to an interesting conclusion: Planting trees, greening unused or vacant spaces, repairing broken windows, and cleaning up litter in the neighborhood may not just create a greener city; it may also create a safer (and healthier) city as well! And for the corollary, removing green space in cities could run the risk of invoking the adverse effects of the broken window theory, so protecting our urban trees and gardens takes on an additional facet of importance.

One last note here: Public perception of what constitutes a "broken win-

---

195 Wikipedia, "History of the New York City Subway."

196 Kate Barnes, "Study: Vacant Lot Greening Can Reduce Community Crime, Violence," University of Michigan, Michigan News (blog), October 28, 2022, https://news.umich.edu/study-vacant-lot-greening-can-reduce-community-crime-violence/.

197 Michelle C. Kondo, SeungHoon Han, Geoffrey H. Donovan, and John M. MacDonald, "The Association between Urban Trees and Crime: Evidence from the Spread of the Emerald Ash Borer in Cincinnati," *Landscape and Urban Planning* 157 (2017): 193-199, https://doi.org/10.1016/j.landurbplan.2016.07.003.

dow" often changes based on the community's ever-changing cultural values. For example, depending on the community, things like visible homelessness or open-air drug usage could lead to perceptions of reduced safety, such as what public transit operators reported earlier in this chapter in explaining why ridership had decreased.*

The topic of homelessness certainly became one of the top agenda items for voters in Seattle in the early 2020s, where I lived for several years. I can personally attest that riding the light-rail next to someone doing fentanyl is not a pleasant experience, though this did not stop me from continuing to use the light-rail.

Given this, it's important to be fluid in the policy positions and priorities we take. As better data emerges and new safety issues crop up, this could lead to an entirely new set of recommendations to drive up city safety to keep residents from moving away.

Discussion Points

1) What has led to perceptions as to the relative safety of your local public transportation options? Have you ever felt unsafe using public transportation? Or been deterred from using it? What could solve this?

2) Where in your community do you think would benefit most from "greening" a public or private space?

Other Miscellaneous Policies

Beyond the above, there are many, many more types of policies that can indirectly help the environment, which would take too long to fully list out here. Here is a quick list of examples that fall into this last catchall category.

- **Smoking Bans:** As mentioned in part 2, cigarette butts are one of the most commonly found types of ocean litter. In addition to creating a healthier society, banning smoking would ban the creation of cigarettes themselves, reducing a significant source of nonbiodegradable litter.

- **Cryptocurrency Regulation:** In part 2, I shared how intensely energy-hungry Bitcoin is as a currency. Policies that would force more regulatory oversight of cryptocurrency or demand energy efficiency

WHAT WE CAN DO

targets as a license to operate (e.g., shifting to proof of stake from proof of work) would be a great place to start.

- **Four-Day Workweeks:** Not only is this position attractive from a work-life balance perspective, but it *could* also lead to reduced emissions by cutting out one day of commuting a week from much of the population. That is, as long as people don't take advantage of the longer weekends by doing emissions-intense activities, like more frequent road trips.

There are plenty more examples like the ones above. Hopefully, by using the framework that has been established in this book, you will identify these and other types of interesting "sustainable under the surface" policies to support!

Discussion Points

1) Are there any policies under active consideration in your community that could benefit the climate? How could you think further about these policies after reading this section?

2) What other policy positions could help generate positive environmental impact?

Chapter 16: Engaging Civically and Running for Office

With all that detail about policy out of the way, you may find yourself frustrated by your slate of local and national politicians. Maybe they are tackling some of these issues but are not prioritizing enough of them (or at least the ones that you are most passionate about). At worst, they might be actively opposing some of them!

Thankfully, even if your local politicians are not prioritizing your preferred policy, your governmental bodies have a mechanism to do so: **public comment**.

Every piece of legislation goes through a period of public comment, where anyone can add their voice to the issue (usually experts and affected community members). This period of public comment can be one mechanism to help your voice be heard on very specific pieces of legislation, which has the potential to positively influence them. (This is also a process that applies to certain jobs and corporations.) It's a great way for those who will be impacted by a new law to weigh in on how it could be shaped and implemented. If you want to take the fight directly to the politicians, you also have the option to vocalize your policy concerns in town halls (if a politician holds them) or by calling or mailing them.

If these levers don't feel sufficient, this is when we must remember that, as citizens, we don't just have the power to vote for or influence our preferred candidates. We also have the power to become candidates ourselves! This is certainly the hardest to do at the national and state political level (I would guess the average reader of this book is unlikely to become a future president or state governor). But given that we live in a democracy, there are far more elections taking place in many aspects of your life that you could throw your hat into the ring for: homeowners associations (HOAs), school boards, mayor, city council, or even a local company or nonprofit board of directors. Personally, I have sat on the board of a nonprofit organization and found it immensely gratifying—and a way to create impact outside of my normal day-to-day.

If you believe you have a worthy cause, then you should work on and support voter access and outreach (e.g., vote by mail, voter participation). One potential added benefit of this work is that, even beyond the immediate campaign you might be running, the causes you are championing could stick with some of the people you canvas. Running for a position on your local HOA board to allow for more "natural" lawns, for example, could help the people in your neighborhood think more holistically about the benefits of biodiversity in their daily lives. Aiming to join your local school board could help influence curriculum and extracurriculars to have more climate-friendly content. And if you're still in school, running for student government can add greening the campus to the school's agenda.

Of course, any sort of "bigger" office you might hold at a city, state, or federal level gives you much broader opportunities—and climate-friendly policies can (hopefully soon!) become a deciding, bipartisan factor as to who can win said elections.

What about Climate Activism and Protesting?

The reader might be a bit confused as to why I've waited until nearly the end of the book to bring up the role that traditional activism and protesting should play in how we engage civically—especially when popular activists like Greta Thunberg (who has no qualms about protesting and getting arrested for her stances) have taken so much global mindshare in the global climate change discourse. Isn't participating in a climate march or protesting a local polluter something worth doing for the planet?

Whether or not protesting and activism are effective is an incredibly nuanced topic with bodies of study that span far beyond just the topic of sustainability (and, though I have participated in a number of protests throughout the years, I will not aim to claim deep expertise on activism myself). Protests are often incredibly context-dependent as well, ranging from what is being protested, to the nature of the protest, to the audience of the protest, to the makeup of the protesters. One group's incredibly successful protest could become another group's incredibly disastrous one.

I think of protesting and activism, in some ways, as tactics—one means to an end. The "end" could be enabling any of the personal, professional, or political outcomes that we've reviewed throughout this book. In some cases, the tactics mentioned within this book *are* activism, even if not in the classical sense of the word nor explicitly billed as such. For the "means" of activism to be successful, I would offer a few guideposts for consideration in line with much of the rest of this book:

1) Is the activism clear on what it aims to achieve with clear and actionable objectives? Or is it too genericized to spur meaningful action?

2) Is the activism directed at the right people who have power to make a difference? Or is it misdirected at people who have no authority to achieve what you are asking for?

3) Does the activism bring others on the journey and allow for them to take steps forward to become involved? Or does it demand immediate perfection, polarizing the issue and driving potential allies away from participating in a climate journey?

4) Does the act of activism itself come with too heavy of an environmental burden? For example, is it "worth it" from an environmental point of view to incur the emissions of a cross-country flight to join a climate march?

All in all, as long as the activism or the protest is intentional and strategic in thinking through the above guidelines, it can (hopefully) avoid some of the pitfalls that can sink an otherwise well-intentioned effort.

Discussion Points

1) Look up some of the public comment opportunities in your town or city. What is open for comment? How might your comments help add sustainability to the mix?

2) If you were to run for a local office, what office would you run for? And what would be your sustainability agenda?

3) What do you think makes for effective versus ineffective climate activism? Can you think of real-world examples that have worked well (or not so well)?

Closing Part Three

In concluding this part on what we can do in our political lives, I hope you have recognized that **just like every job can be a sustainability job, every vote can be a sustainability vote**. There are a lot of powerful sustainability (and sustainability-adjacent) policies and activities that are worth supporting, sometimes in surprising ways. And engaging with nonprofits and local organizations can be an excellent way to create a real and tangible impact on sustainability in your immediate communities.

Hopefully, you have started to see the through line that can connect impact across your personal, professional, and political lives. Voting for a politician who supports renewable energy access could make it easier to convince your employer to put solar panels on their roof—and for you to do the same for your home. Successfully advocating for Right to Repair in a state town hall could create an invitation for you to help advise on repairable design opportunities for your company's products, letting you extend the lifespan (and value) of those products for longer in your home. Getting engaged with a local nonprofit could transform into a partnership between it and your employer to help clean up your local park, which you could enjoy as a local place to enjoy greenery on the weekends.

Before we head off to the conclusion of this book, it's worth stressing that our communities are only as strong as the people within them. If your community doesn't have a climate advocate, consider becoming its standard bearer. Your leadership could influence others within your community to join you on a sustainability journey. If they, too, can reconsider how to embed sustainability into their lives, their work, and their community, then it's only a matter of time before the multiplicative effect could lead to real and lasting change. To restate from the introduction of this book, the impact of the many exceeds the impact of the few.

Discussion Points

1) What sustainability issues do you think could best galvanize your community to take action on the environment? What might prevent them from getting involved?

2) What are the top things you could do in your community to make a difference today?

3) How will you vote in your upcoming election(s) to maximize your community's potential for sustainable action? What are the main drivers of impact at stake?

CONCLUSION

To close out this book, I want you to think back to a question posed in the introduction: "How can we be sure we're actually doing the best things for the planet?"

If this book has been successful, you've been taking notes along the way about tidbits that you have learned and hope to apply in your life going forward. You're seeing things as a holistic mix of the emissions associated with the "Make It, Move It, Use It, Lose It" framework. Maybe you've even had a book club to walk through some of the discussion points. I know that I learned a lot while researching this book, and I have changed my behavior in interesting ways to align with the latest science. For example, now that I know that heating water is one of the biggest sources of household emissions in America, I have shifted from always rinsing my dishes in warm water to using cold water instead. I've also realized the added importance of the seasonality of the ingredients I use in my cooking and am trying to be better at "cooking with the seasons."

If I were to sum up the immense complexity of living in modern society and translate it into the "Top Ten Things We All Can Do to Make the Biggest Impact" based on this book's chapters, I think the list would have to look something like what's below. Acknowledging that some levers will be more accessible (and more impactful) for some people than others, this list is ranked in order of quantified (or at least estimated) impact:

1) **Bring Others Along on the Journey:** Ultimately, our impact is most pronounced when others can amplify it. Remember our rallying cry from the beginning of the book: "The impact of the many exceeds the impact of the few." More allies mean that each of us can go on our own journey to embed sustainability into our personal, professional, and political lives.

2) **Get Civically Engaged and Vote:** Exercising your right to vote—from federal and state elections to local school boards—can drive the most powerful lever that we as individuals ultimately have: influence on policy, laws, and regulations. Getting civically engaged, whether by advocating for the right causes or even running for office yourself, doesn't guarantee change immediately. But when it does, the change potential can be massive.

3) **Make Your Job a Sustainability Job:** Rethinking how our careers can become integrated with sustainability, whether by layering sustainability into our existing jobs or seeking a new career entirely, is something we all can do to extend our impact. After government, companies

are some of the most far-reaching and powerful entities in the country, and driving change at these companies can yield a significant impact.

4) **Put Your Money (and Time) to Work for the Planet:** Donating money or volunteering for environmental (and environmental-adjacent) nonprofits can lead to a significant "return on impact" investment. Further, keeping your savings at sustainable banks and credit unions, using eco-friendly credit and debit cards for spending, and investing in ESG funds (or "provesting" in dirty companies) have the potential to support clean-energy and product transitions.

5) **Be Intentional about Family Planning:** How we grow and plan our families, in aggregate, ultimately determines how many people are around to cause emissions and consume resources. Planning a smaller family, an older family, an adopted family, or even no family at all can help us minimize a lifetime's worth of emissions.

6) **Go Solar—and Go Electric:** Powering our homes (and perhaps our cars!) with renewable energy is more accessible than ever before, and this can significantly reduce the biggest chunk of emissions in most of our daily lives. Switching our heating (and our cars) from being gas-powered to electric-powered can help our homes reduce the rest.

7) **Cut Down on Solo Driving and Long Trips:** Getting around by carpooling or other alternative forms of transport, such as buses or bicycles, can tackle our next biggest source of personal emissions. Switching to hybrid or remote working, if possible, can help ease this transition as well. And when it comes time to take vacations, the less distance you need to travel by car, plane, or boat, the better!

8) **Get Efficient in Your Home:** Winterizing (and "summerizing") your home and using your heating, cooling, appliances, and electronics more efficiently can knock down the amount of energy you use significantly. And don't forget: one of the best ways to get more efficient in your home is by living with other people—and by living in more modest-sized homes.

9) **Improve Your Diet by Reducing Your Red-Meat Intake:** The biggest source of impact from the things we purchase frequently, our diets have a significant environmental impact. If you can't or aren't willing to become vegetarian or vegan, cutting down on your red-meat

intake in favor of other proteins (e.g., chicken), reducing food waste, and buying seasonally and locally can cut the impact of your diet by more than half.

10) **Reuse Your Things, Buy Used, and Buy in Bulk:** For everything else, because of the outsized impact it takes to make things, try reducing the number of things you need, reusing the stuff you already have, and buying used or in bulk when possible. And, of course, if you can't find a second life for your things when you're done with them, always try to recycle if the product allows it.

Remember, this is only a directional guide. Some of us already live lives that may reduce—or increase—the relative impact of some of these categories. In some cases, the ordering may change slightly based on your personal circumstances. And if reducing water usage or waste generation is your preferred way to help the environment, this top ten list would look a little bit different. But overall, starting with our political lives, then our professional lives, and finally our personal lives will give us the best mechanisms to make a meaningful, lasting impact. And rather than feeling pressure to do all ten of these things or chastising others for not doing everything on this list, we should celebrate every time we "turn the dial" toward more sustainable living.

With all this being said, I can completely understand if it still doesn't feel like we can do enough to make a difference. Climate change, global pollution, and water stress can feel like some of the most intractable problems that humanity has ever faced. To many, it feels like an existential threat to the future of all life on our planet—and a near-term catastrophe, as climate change displaces millions in search of water or fleeing from fire or famine.

But just as the physical impacts of climate change are rearing their ugly heads, so too are the societal and psychological impacts. From a Yale study in 2022, about one in ten Americans reported "feeling down, depressed, or hopeless for at least several days out of the last two weeks because of global warming." Around the same number reported experiencing symptoms of depression from climate anxiety, and 14 percent think "it's too late to do anything about global warming."[198]

Let's also look back to Yale's "Global Warming's Six Americas" research

---

[198] Anthony Leiserowitz, Edward Maibach, Seth Rosenthal, John Kotcher, Jennifer Carman, Marija Verner, Sanguk Lee, Matthew Ballew, Sri Saahitya Uppalapati, Eryn Campbell, Teresa Myers, Matthew Goldberg, and Jennifer Marlon, "Climate Change in the American Mind: Beliefs & Attitudes, December 2022," Yale Program on Climate Change Communication, Yale University, February 16, 2023, https://climatecommunication.yale.edu/publications/climate-change-in-the-american-mind-beliefs-attitudes-december-2022/toc/2/.

mentioned in the introduction. Between 2013 and 2023, the number of people who self-reported as "alarmed" about climate change grew from 15 percent to 28 percent—the single biggest jump of all responses.[199]

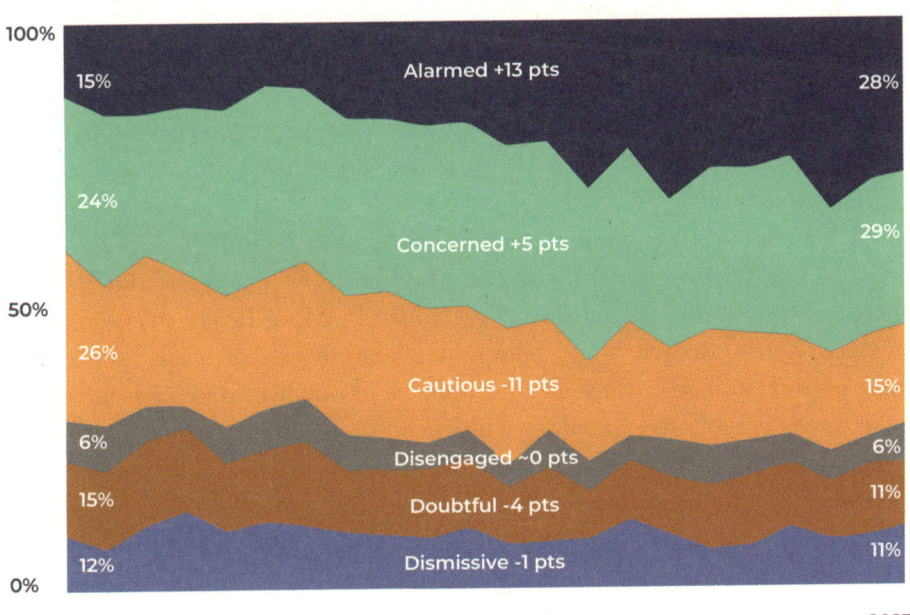

Source: "Global Warming's Six Americas," Yale Program on Climate Change Communication, https://climatecommunication.yale.edu/about/projects/global-warmings-six-americas/.

Other studies affirm this, too, especially for young people between the ages of sixteen and twenty-five. A 2021 study from the University of Bath in England across ten different countries found that, globally, 45 percent of young people said that their feelings about climate change negatively affected their daily life and functioning, and around three-quarters of respondents said they thought that "the future is frightening."[200]

This anxiety is seeping into some of the most personal decisions we make

---

199 "Global Warming's Six Americas," Yale Program on Climate Change Communication.

200 Caroline Hickman, Elizabeth Marks, Panu Pihkala, Susan Clayton, R. Eric Lewandowski, Elouise E. Mayall, Britt Wray, Catriona Mellor, and Lise van Susteren, "Climate Anxiety in Children and Young People and Their Beliefs about Government Responses to Climate Change: A Global Survey," *Lancet Planetary Health* 5, no. 12 (2021): e863–e873, https://doi.org/10.1016/S2542-5196(21)00278-3.

as well. For example, a study from the Pew Research Center in 2021 found that 5 percent of all American adults between the ages of eighteen and forty-nine who have not yet had children cite climate anxiety as their primary reason for not wanting to do so.[201] And for young people specifically in the University of Bath study? About 25 percent indicated climate change as a reason to not have children.[202]

Climate anxiety is leading to climate doomism. I don't want to belittle or cheapen this anxiety—I often feel it too. But I also don't want to shift the "burden of action" squarely on the average person to the point where the responsibility is just as crushing. Large corporations and governments ultimately have the most responsibility to usher us into a zero-emission, zero-waste, and water-positive future.

What I hope has come out of reading this book is, if nothing else, a spark of hope and optimism that we're not as powerless as we might have once felt. There are meaningful frameworks (e.g., "Make It, Move It, Use It, Lose It," "Reduce, Reuse, Recycle") and options available to us that teach us *how* to make more sustainable decisions and impacts in our lives and *where* we can make those decisions to expand our influence.

For the How: We've benefited from a nascent revolution in carbon, waste, and water accounting to realize that *making* things can often be even more wasteful than *using* things. As a result, supporting behaviors toward a more "circular" economy—reducing, then reusing, then recycling—can influence every decision we make for what we make, buy, use, and discard. We have the power to make a sustainability impact in every decision we make.

For the Where: We've realized that the power of the many exceeds the power of the few. Thinking beyond just our individual decisions and toward the jobs, communities, and municipalities that we belong to, we have far more power in our hands to influence more broadly than we might have ever realized. Every job can be a sustainability job, and every vote can be a sustainability vote.

I imagine if you've read to the end of this book, the **why** for making sustainable decisions in our lives is something that you've already thought quite a lot about. Maybe it's because you want to leave a better planet for your children or the next generation. Maybe it's because you want to feel like you're making more of an impact. Or maybe it's because you're looking to reorient your North Star toward a higher purpose.

---

201 Anna Brown, "Growing Share of Childless Adults in U.S. Don't Expect to Ever Have Children," Pew Research Center, November 19, 2021, https://www.pewresearch.org/short-reads/2021/11/19/growing-share-of-childless-adults-in-u-s-dont-expect-to-ever-have-children/.

202 Hickman, Marks, Pihkala, Clayton, Lewandowski, Mayall, Wray, Mellor, and van Susteren, "Climate Anxiety in Children," e863–e873.

After all this, the only things left to answer are the **who** and the **when.**
The who is you.
The when is now.

Discussion Points

1) Do you experience climate anxiety? If so, how has it manifested in how you see the world and make decisions?

2) Reflecting on the lessons you've learned across the book, what stood out to you most? How could you apply that lesson going forward?

3) What are the three things you are planning to do differently going forward to lead a more sustainable life? Have you started doing anything differently already?

APPENDIX: SELECT DATA AND DEFINITIONS

WHAT WE CAN DO

In chapter 1, the opening pie chart was presented as a view into the emissions profile of an "average" American across their daily lives.

The following are the assumptions made for each portion of this pie, as well as the calculations made and data sources used. Note that calculations are meant to provide directional guidance and are not an attempt to derive a perfectly scientific analysis of emissions estimation.

Category	Assumptions
Heating	Assumes 2.5 people per household.
Cooling	
Water Heating	
Refrigerators and Freezers	
Lighting	
Clothes Dryers	Given the known amount of emissions for water heating from the above, uses ratio of British thermal units (BTU) recorded by the EIA for each respective category against water heating to estimate the impact of categories embedded in the "Other" category of the EPA data analysis.
Cooking and Kitchens	
Electronics Usage	
Other Home Emissions	Assumes lawn care and pet care are excluded from home emissions in the EPA study; subtracts known emissions from the above categories to determine residual emissions not accounted for in the above categories.
Lawn	Assumes 80 percent of households have a lawn.
Pets	Assumes the household has one pet, a midsized dog, and no other pets. Assumes 2.5 people per household.
Household Embodied Carbon	Assumes a 50-year lifespan for a house, 2.5 people per household, household size of 200m², and full cradle-to-grave emissions.
Commuting	Assumes the commuter is driving solo, and one commuter per household.
Household Driving	Assumes the car is shared among 2.5 people within the household for non-commuting driving.
Vacations	Uses two round-trip flights from Los Angeles to New York City as a proxy for generalized vacation travel.
Vehicle Embodied Carbon	Assumes 11-year vehicle lifespan.
Food	Assumes average diet by dividing total agriculture emissions by the population of America.
Electronics Embodied Carbon	Assumes 1 new electronic purchased per year. Uses Surface Laptop Go 3 as a proxy for embodied carbon.
Clothes	Uses jeans as a proxy for all clothing purchase types. Assumes 10 clothing purchases a year.
Furniture	Assumes 2 new pieces of furniture purchased annually.
Appliances Embodied Carbon	Uses 2 refrigerators as a proxy to represent multiple appliances. Assumes an average refrigerator lifespan of 14 years.
Other	Rough estimate as a proxy stand-in. Assumes 5 books purchased a year (3 kg/book), 50 rolls of toilet paper (0.5 kg/roll), and value doubled to cover other purchases.

WHAT WE CAN DO

Calculation	Data Source
[Total emissions of a household] x [% for heating] / [Average people in the household]	See footnotes 203 and 204.
[Total emissions of a household] x [% for cooling] / [Average people in the household]	See footnotes 203 and 204.
[Total emissions of a household] x [% for water heating] / [Average people in the household]	See footnotes 203 and 204.
[Total emissions of a household] x [% for refrigeration] / [Average people in the household]	See footnotes 203 and 204.
[Total emissions of a household] x [% for lighting] / [Average people in the household]	See footnotes 203 and 204.
[Sum of BTU for washers and dryers] / [Sum of BTU for water heating] x Water heating emissions	See footnote 205.
[Sum of BTU for cooking, microwaves, dishwashers] / [Sum of BTU for water heating] x Water heating emissions	See footnote 205.
[Sum of BTU for TVs and related] / [Sum of BTU for water heating] x Water heating emissions	See footnote 205.
[Total emissions of a household] / [Average people in the household] − [Sum of all known home emissions sources]	See footnote 206.
[Lawn care emissions in America] / [Population of America] / [% of people with lawns]	See footnote 207.
[Average lifetime emissions of a dog / Life expectancy of a dog] / [Average people in household]	See footnote 208.
[Total embodied carbon of a household] / [Lifespan of house] / [Average people per household]	See footnote 209.
[EPA emissions from 1 year of driving average vehicle] x [% of trips for commuting]	See footnotes 210 and 211.
[EPA emissions from 1 year of driving average vehicle] x [% of non-commuting trips] / Average people per household	See footnotes 210 and 211.
[Emissions taken from Google Flights carbon emissions estimator]	See footnote 212.
[Gas vehicle manufacturing emissions] / [Estimated vehicle lifespan] / [Average people per household]	See footnote 213.
[Total emissions from agriculture industry] / [Population of America]	See footnote 214.
[Manufacturing + distribution emissions for proxy electronic] x [Number of electronics purchased per year]	See footnote 215.
[Sum of non-usage emissions from Levi's jeans] x [Number of clothing purchased per year]	See footnote 216.
[Estimated emissions of average furniture] x [Number of furniture purchases per year]	See footnote 217.
[% of life-cycle emissions from manufacturing for a refrigerator] x [Lifetime expectancy of a fridge] x [Refrigerator annual usage emissions] / [People in Household] x [Number of appliances]	See footnote 218.
[Number of books purchased per year] x [Book emissions] + [Number of toilet paper purchased per year] x [Toilet paper emissions] x 2	Miscellaneous

239

Part 1 also generated an estimate of per-household emissions for each state within America. Using data from the EIA[219] and cross-referenced with each state's population per the 2020 US Census, this yielded the following ranked list of annual residential carbon emissions per capita in metric tons of carbon dioxide equivalent ($mtCO_2e$):

---

203 "Use of Energy Explained," US Energy Information Administration.

204 United States Environmental Protection Agency, "Greenhouse Gases Equivalencies Calculator—Calculations and References."

205 US Energy Information Administration, "Table CE5.1a."

206 "Use of Energy Explained," US Energy Information Administration.

207 Banks and McConnell, "National Emissions from Lawn and Garden Equipment."

208 Bottollier-Depois, "Carbon Pawprint."

209 Magwood, Huynh, and Olgyay, "The Hidden Climate Impact."

210 National Household Travel Survey, "Popular Vehicle Trips Statistics."

211 United States Environmental Protection Agency, "Greenhouse Gas Equivalencies Calculator."

212 Google, "Google Flights."

213 "Electric Vehicle Myths, Footnote 2," United States Environmental Protection Agency, last modified June 11, 2024, https://www.epa.gov/greenvehicles/electric-vehicle-myths.

214 United States Environmental Protection Agency, *Inventory of U.S. Greenhouse Gas Emissions and Sinks.*

215 Microsoft, "Eco Profiles."

216 Levi Strauss & Co., *The Life Cycle of a Jean*, slide 1.

217 Furniture Industry Research Association, *A Study into the Feasibility.*

218 LG Electronics, "Products Application."

219 US Energy Information Administration, "Energy-Related CO2 Emission Data Tables."

WHAT WE CAN DO

State	Annual Residential Energy Emissions (M mtCO2e)	Population as of the 2020 Census	Annual Residential Energy Emissions per Person (mtCO2e)
Florida	1.4	21,538,187	0.07
Hawaii	0.1	1,455,271	0.07
Arizona	2.6	7,151,502	0.36
South Carolina	1.9	5,118,425	0.37
Alabama	1.9	5,024,279	0.38
Louisiana	1.9	4,657,757	0.41
Texas	12.1	29,145,505	0.42
North Carolina	4.8	10,439,388	0.46
Mississippi	1.4	2,961,279	0.47
Tennessee	4	6,910,840	0.58
Arkansas	1.9	3,011,524	0.63
Oregon	2.8	4,237,256	0.66
Georgia	7.1	10,711,908	0.66
Kentucky	3	4,505,836	0.67
Virginia	5.8	8,631,393	0.67
California	26.6	39,538,223	0.67
Washington	5.8	7,705,281	0.75
Nevada	2.7	3,104,614	0.87

State	Annual Residential Energy Emissions (M mtCO2e)	Population as of the 2020 Census	Annual Residential Energy Emissions per Person (mtCO2e)
District of Columbia	0.6	689,545	0.87
Maryland	5.5	6,177,224	0.89
Delaware	0.9	989,948	0.91
West Virginia	1.7	1,793,716	0.95
Oklahoma	3.8	3,959,353	0.96
Missouri	6.4	6,154,913	1.04
New Mexico	2.3	2,117,522	1.09
Idaho	2	1,839,106	1.09
Indiana	8.3	6,785,528	1.22
South Dakota	1.1	886,667	1.24
Nebraska	2.5	1,961,504	1.27
North Dakota	1	779,094	1.28
Utah	4.2	3,271,616	1.28
Kansas	4	2,937,880	1.36
Pennsylvania	17.8	13,002,700	1.37
Ohio	17.1	11,799,448	1.45
Colorado	8.5	5,773,714	1.47
New Jersey	14	9,288,994	1.51

WHAT WE CAN DO

State	Annual Residential Emissions (mtCO2e)	Population as of the 2020 Census	Annual Residential Carbon Emissions per Person (mtCO2e)
New York	31.5	20,201,249	1.56
Wyoming	0.9	576,851	1.56
Minnesota	9.2	5,706,494	1.61
Iowa	5.2	3,190,369	1.63
Montana	1.8	1,084,225	1.66
Wisconsin	9.8	5,893,718	1.66
Rhode Island	1.9	1,097,379	1.73
Massachusetts	12.2	7,029,917	1.74
Illinois	23.4	12,812,508	1.83
Connecticut	6.6	3,605,944	1.83
Michigan	19.6	10,077,331	1.94
New Hampshire	2.8	1,377,529	2.03
Maine	2.9	1,362,359	2.13
Vermont	1.4	643,077	2.18
Alaska	1.7	733,391	2.32

Glossary

Carbon Accounting: The application of measurement to the emissions associated with various activities. This includes methods such as the lifecycle assessment.

Carbon Dioxide Equivalent (CO_2e): A metric that combines the global warming potential of many different types of emissions (e.g., methane, carbon dioxide, nitrous oxide) into a single measurement indexed to the global warming potential of carbon dioxide. For example, methane is roughly twenty-eight times as potent as carbon dioxide, so 1 kilogram of methane is equal to 28 kilograms of CO_2e.[220]

Carbon Tax: A tax levied based on the amount of emissions generated by an organization. The higher the emissions, the higher the tax the emitter must pay.

Circular Economy: In contrast to a "linear economy," where we manufacture things from virgin raw materials and throw things out after we use them, this term refers to an economic structure that incentivizes the use of recycled materials in production as well as the reuse of products by consumers. An advanced form of "Reduce, Reuse, Recycle."

Climate Change: In this book, this refers to the anthropogenic (i.e., human-caused) change of Earth's climate, generally resulting in a warmer planet with more unpredictable weather patterns than Earth would itself naturally produce.

Climate Doomism: The general mentality that we are powerless to prevent or slow down climate change and therefore "doom" is inevitable. This is often tied to feelings of despair, nihilism, and/or anxiety.

Climate Optimism: The general mentality that climate change is a solvable

---

220 United States Environmental Protection Agency, *Inventory of U.S. Greenhouse Gas Emissions and Sinks*.

problem and that we have a tangible role to play in solving it. Ideally, the outcome of this book.

CO_2: Shorthand for carbon dioxide.

Embodied Carbon: Emissions that were generated to manufacture a thing rather than to use or transport it.

Extended Producer Responsibility: A concept wherein companies are held accountable for the externalities associated with their products.

Externality: A cost of doing business that is shifted from a company onto the general population, such as how recycling is paid for by tax dollars rather than by the companies producing the products that need to be recycled, or how pollution from a company can incur higher medical bills that are paid for by the individuals subjected to it.

Global Warming: A consequence of greenhouse gas emissions that are raising the average temperature of Earth. This is one way in which we are experiencing climate change.

Greenhouse Gas: Any gas such as methane or carbon dioxide that if released into the atmosphere will lead to the warming of Earth by trapping heat (hence, the greenhouse effect).

Greenwashing: The general practice of an entity (typically a company) using misleading or false statements to overstate the sustainability of a product or activity.

Lifecycle Assessment: A bottom-up type of carbon accounting that accounts for the full lifetime emissions of a single thing or product, inclusive of the supply chain impacts that specifically contributed to the creation of the product as well as the estimated impact of its usage by consumers of the product.

Low-Carbon Energy: Energy sources that do not create a significant amount of greenhouse gas emissions in their production. This includes renewable energy as well as nuclear energy.

Miles per Gallon Equivalent (MPGe): Typically used to describe how far an electric vehicle could travel using the same amount of energy as is found in 1 gallon of gasoline.

Per-Capita Emissions: The allocation of the emissions generated by a shared activity (e.g., a car pool) to each individual participating in the activity. For example, the per-capita emissions of two people in a car pool are one-half of the total emissions of the car.

Power Purchase Agreement (PPA): A contract in which the buyer agrees to purchase energy at a fixed price from the seller over a given period. These can specify the type of energy purchased and are often used to finance renewable energy projects.

Provestment: In contrast to "divestment," provestment is an impact strategy focused on investing in "dirty" companies to gain increased voting power for shareholder proposals focused on driving positive environmental and social change.

Recycle: The general practice of preventing an item from going to landfills by returning it (most often to a municipality) so that its raw materials can be reclaimed for new use.

Reduce: The general practice of consuming fewer things, as expressed by the purchase of fewer things or the decreased usage of a purchased item.

Renewable Energy: Energy that does not come from a finite source. Wind, solar, and hydroelectric power are considered renewable. Nuclear, oil, gas, and coal are considered nonrenewable. Renewable energy tends to produce little to no emissions.

Reuse: The general practice of using a thing as long as possible over many uses, repairing it rather than discarding it or giving it a second life (e.g., by purchasing refurbished or by giving the item to a new owner).

Scope 1 Emissions: Most often refers to on-site emissions from the burning of fuel, such as driving a company shuttle or using gas to heat the office.

Scope 2 Emissions: Refers to the emissions generated by the use of electricity on company property, such as to power lighting and air conditioning or to run computer servers.

Scope 3 Emissions: Refers to the emissions generated by others on behalf of the company, both the suppliers to the company as well as the customers of the products.

Supply Chain: The collection of activities required to manufacture and deliver a product to the customer. The "Make It" and "Move It" parts of the "Make It, Move It, Use It, Lose It" framework.

Watts and Kilowatts: Simply, a watt measures the rate at which an electricity-powered device uses power. A kilowatt is equal to 1,000 watts. A kilowatt-hour (kWh) describes the amount of energy that 1,000 watts would use in one hour.

About the Author

Charlie Sellars is a director of sustainability at Microsoft, which pledged to become carbon negative, water positive, zero waste, and protect ecosystems by 2030. As one of the youngest directors at the company, he has overseen sustainability for both the Windows and Devices and Cloud Operations portions of Microsoft, helping launch several sustainability-forward products ranging from new Windows PCs with repairable and recycled components to the Ocean Plastic Mouse. Recognized by IM100 as one of 2024's top 100 most impactful individuals in the digital infrastructure industry, Charlie also serves as a governing body member of the iMasons Climate Accord, an industry coalition united to decarbonize the digital infrastructure that underpins the next generation of cloud and AI services.

Charlie has previously served as board member and CTO of an impact-focused non-profit, The $100 Solution, which believes that "solutions to big problems start with small steps." He initially joined this non-profit while studying for his Bachelor of Arts degree in physics from Williams College, a small liberal arts school nestled in the Berkshire mountains which helped to grow his love for nature.

Charlie's unique approach uses data-driven steps and daring optimism to empower all readers to maximize sustainability in their personal, professional, and political lives.

For more resources, find Charlie online at www.charliesellars.com.